Managing construction projects

MANAGING CONSTRUCTION PROJECTS

A GUIDE TO PROCESSES AND PROCEDURES

Edited by A. D. Austen and R. H. Neale

International Labour Office   Geneva

Copyright © International Labour Organisation 1984

Publications of the International Labour Office enjoy copyright under Protocol 2 of the Universal Copyright Convention. Nevertheless, short excerpts from them may be reproduced without authorisation, on condition that the source is indicated. For rights of reproduction or translation, application should be made to the Publications Branch (Rights and Permissions), International Labour Office, CH-1211 Geneva 22, Switzerland. The International Labour Office welcomes such applications.

ISBN 92-2-106476-X

First published 1984
Third impression 1990

The designations employed in ILO publications, which are in conformity with United Nations practice, and the presentation of material therein do not imply the expression of any opinion whatsoever on the part of the International Labour Office concerning the legal status of any country, area or territory or of its authorities, or concerning the delimitation of its frontiers.
The responsibility for opinions expressed in signed articles, studies and other contributions rests solely with their authors, and publication does not constitute an endorsement by the International Labour Office of the opinions expressed in them.
Reference to names of firms and commercial products and processes does not imply their endorsement by the International Labour Office, and any failure to mention a particular firm, commercial product or process is not a sign of disapproval.

ILO publications can be obtained through major booksellers or ILO local offices in many countries, or direct from ILO Publications, International Labour Office, CH-1211 Geneva 22, Switzerland. A catalogue or list of new publications will be sent free of charge from the above address.

PREFACE

This Guide describes the processes and procedures of construction project management, with emphasis on their use in developing countries. Good project management is essential because of the importance of capital projects to the development of a young nation. In many developing countries construction alone accounts for about 10 per cent of the gross national product, and 50 per cent or more of the wealth invested in fixed assets. The importance of construction work in providing the physical facilities for development activities is indicated in figure 1.

The Guide is not intended as a manual; it does, however, describe the general principles of construction project management, and it emphasises many of the vital

Figure 1. Construction work in developing countries.

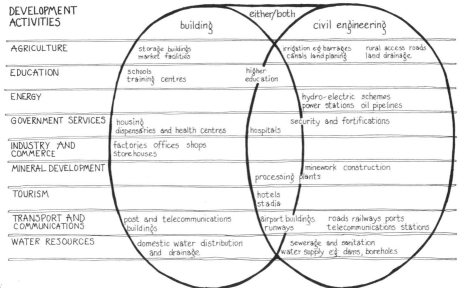

requirements for the successful execution and completion of projects. There is especial emphasis on the need for careful planning in the early stages of projects, as well as during the activities intended to prepare the users to take over, use and maintain the facilities provided.

The conditions in which construction projects are undertaken differ from one country to another, but the principles advocated herein are of a general character and are based on internationally accepted procedures.

Figure 1 also shows that the word "construction" covers both building and civil engineering work and that some construction projects, notably complex buildings, may include both. Many of the principles described apply to all construction projects; but where the processes and procedures relevant to building and civil engineering work differ significantly, they have been given separate explanations.

ACKNOWLEDGEMENTS

This Guide has been developed as part of a technical co-operation programme between the Swedish International Development Authority (SIDA) and the International Labour Office. A preliminary edition was used in seminars and training courses in developing countries. Experience gained thereby resulted in the present edition, which covers the subject more comprehensively and gives some separate treatment to building and civil engineering work.

The preliminary edition of the Guide was a joint effort by a number of people: Leif Lindstrand and Sigurd Stäbgren, National Swedish Board of Building; Derek Miles, Intermediate Technology Development Group, London; John Andrews, University College, London; Bengt Göransson, Hifab International AB, Stockholm; Mikael Söderbäck, formerly Hifab International AB, now SIDA; John Brisman and Nils Nilsson, White Arkitekter, Stockholm.

The present edition was prepared and edited by A. D. Austen, of the ILO Management Development Branch, and R. H. Neale, Senior Lecturer, Department of Civil Engineering. Loughborough University of Technology, United Kingdom.

Thanks are also due to ILO colleagues and construction management experts whose advice and comments were helpful in the preparation of this Guide.

CONTENTS

Preface i

Acknowledgements vii

1. Introduction 1

2. A building project 5
   Participants 5
   Stages and aspects 7
   Briefing stage 10
   Designing stage 12
   Tendering stage 14
   Constructing stage 17
   Commissioning stage 18

3. A civil engineering project 21
   Characteristics of civil engineering projects 21
   Participants 23
   Stages and aspects 24
   Briefing or report stage 25
   Designing stage 27
   Tendering stage 29
   Constructing stage 30
   Commissioning stage 32

4. Organisation of management functions 35
   Objectives 35
   Project management team 35
     Team functions 36
     Team organisation 36
     Team members: building projects 39
     Team members: civil engineering projects 45
     The reality of management 47

5. Planning 49
   Participants 49
   Principles 50
   Techniques 50
   Activities 52

6. Procurement 59
   Objectives 59
   Participants 60

Methods   60
   Appointing consultants   64
   Appointing contractors   66
   Appointing subcontractors   70
   Appointing suppliers   71

7. Control   73
   Objectives   73
   The role of the project manager   74
   Time, cost and quality control   75
      Time control   75
      Cost control   78
      Quality control   85
      Site control   86

8. Health and safety   93
   Objectives   93
   Participants   93
   Principal factors   93
   Activities   96
   Causes of accidents   99
   Project management team functions   100

9. Communication and reporting   101
   Managing people   101
      Informal communications   103
      Formal communications   104
      Striking the right balance   105
   Meetings   105
      Planning the meeting   106
      Conduct of the meeting   106
      Follow-through   107
   Summary   107

10. Planning techniques   109
    Time-chainage charts   110
    Network analysis   113
       "Activity-on-node" or "precedence" method   113
       "Activity-on-arrow" method   123
       Network analysis in practice   125
    Bar or Gantt charts   126

Appendices

A. Checklists   131
   Briefing stage   131
   Designing stage   133
   Tendering stage   135
   Constructing stage   136

        Commissioning stage   138
        Project administration   140

    B. Specimen job description for project manager   143

    C. Glossary   145

    D. Select bibliography   157

FIGURES

1. Construction work in developing countries.   v
2. The elements of project management.   1
3. The managerial cycle.   2
4. Stages of a construction project.   3
5. The project management team.   4
6. Aspects of a construction project.   8
7. Participants at each stage of a building project.   9
8. Relationship between design influence and cost.   10
9. Design-stage activities for building projects.   13
10. Participants at each stage of a civil engineering project.   25
11. The briefing or report stage.   27
12. Design-stage activities for civil engineering projects.   28
13. Steering committee and project management team.   38
14. Progress of a civil engineering project based on a possible water supply scheme.   46
15. Planning throughout the project.   49
16. A simple network planning diagram.   52
17. Requirements for funds.   56
18. Procurement: the acquisition of project resources.   59
19. The standard approach.   61
20. The early selection approach.   62
21. The design-and-construct approach.   62
22. The divided contract approach.   63
23. The selection of consultants.   65
24. The tendering process.   66
25. Control means monitoring progress and taking appropriate action.   73
26. The control cycle.   74
27. Outline control plan.   76
28. A construction programme.   76
29. The project budget.   78
30. As estimating accuracy increases, deviation from "goal" decreases.   81

31. The structure of cost monitoring and prediction. 82
32. Detailed bar chart. 83
33. "S" curve showing planned and actual values. 84
34. "S" curves, applications and cash payments. 84
35. Quality control throughout the project. 85
36. Use of control plan. 86
37. A contractor's control curve showing the need for working capital. 91
38. Principal factors in effective health and safety management. 94
39. Occupational health and safety and the completed works. 96
40. Formal communications work through the organisational hierarchy. 102
41. Formal and informal communications. 103
42. Face-to-face communication. 104
43. The selection of planning techniques. 109
44. Construction of a simple earth road. 110
45. Plan and section of a simple earth road. 111
46. Time-chainage chart for road shown in figure 45. 112
47. The Kenyan low cost modular timber bridge. 113
48. Cross-section of typical bridge. 114
49. Erection of a bridge. 114
50. Network diagram for bridge using "activity-on-node" or "precedence" method. (The diagram has been divided into two parts to make it fit the page; activities 9 and 12 are repeated to link the parts.) 116
51. Network diagram for bridge using "activity-on-arrow" method. 124
52. Bar chart from figure 50. 127

INTRODUCTION 1

Managing a project is quite different from managing a "steady-state" organisation. A project has a distinct beginning and end, whereas steady-state organisations run continuously. Examples of the latter are hospitals and mass-production factories. In a hospital the basis of medical care changes slowly, despite technological advances, and the administration works to a routine. In mass-production industries the routine of production proceeds continuously, except when new models are being introduced.

Figure 2 shows the main elements of the project man-

Figure 2. The elements of project management.

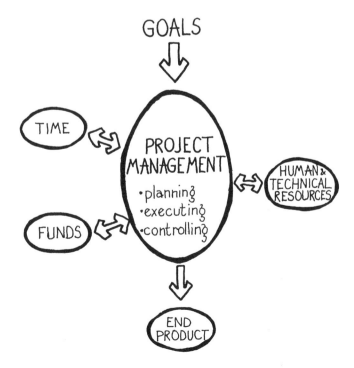

agement process which result in an end product. The cycle of activities to achieve the project goals is shown in figure 3. Since changes, often unforeseen, occur during the lifetime of a project, figure 3 represents continuous action aimed at achieving the best possible result.

Because the construction industry supplies the basic requirements of shelter, water, sanitation, roads, schools and hospitals, its performance has a marked effect both on the economy and on social conditions. This is especially true in developing countries, where much of this infrastructure is lacking. It follows that the efficient management of construction projects is vital if scarce resources are not to be wasted. This Guide seeks to improve managerial effectiveness by describing the way in which a construction project works and how it can be controlled. It does not claim to describe ideal solutions, because these rarely exist in the real world; but it offers a rational and logical approach to the management of construction projects.

Figure 3. The managerial cycle.

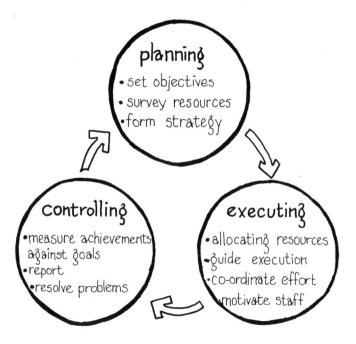

The Guide covers the stages from the time when a decision is made to implement a construction project until the project becomes a reality in the form of a house or a road that has to be maintained. It does not describe the social, economic and engineering analyses which precede or result in that decision. The five stages used in the Guide are shown in figure 4. Their scope may differ from one project to another, but their content should be clearly described for each project.

The main parties involved in a construction project are—

- ☐ the client;
- ☐ the users;
- ☐ the designers;
- ☐ the executors;
- ☐ public authorities and agencies.

The link between them is often provided by a project management team created for the duration of the project (see figure 5). Such a team is unlikely to be a static or permanent body. Its membership may change during the course of the project. At any one stage it should include all persons or parties involved in the work at that stage, such as designers and specialists, user representatives, contractors, suppliers and managerial staff. It will normally be headed by a project manager, who is responsible to the client for the execution of the project. The Guide describes how this way of managing a construction project works, with particular reference to developing countries. It is hoped that it will help those who manage construction programmes to evolve solutions suited to their natural circumstances.

A glossary of terms used in this book appears as Appendix C.

Figure 4. Stages of a construction project.

Figure 5. The project management team.

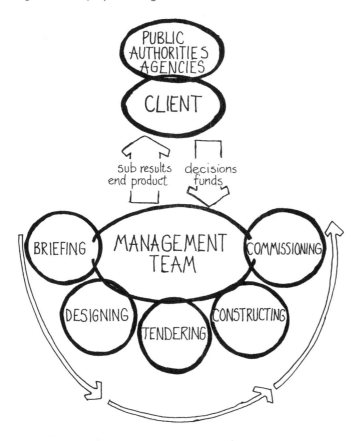

A BUILDING PROJECT 2

participants

The five major groups of people (client, users, designers, executors, public authorities and agencies) participating in a building project can be defined as follows.

THE CLIENT

The client may be an individual, such as someone wishing to build a house. The word is more generally used for the organisation which needs the end product and has the authority (and the money) to order and approve it. For government projects, the client is usually a ministry or department. For example, a Ministry of Health wishing to build a number of health centres may be the "client" of the Ministry of Works. In the private sector the client is often a company. A manufacturing company wishing to build a new factory is the client of the architects it employs to design it and to supervise construction. When a ministry or a company employs a firm of architects, it must pay a fee, so that it is natural for the firm to regard the ministry as its client. A Ministry of Works undertaking work for another ministry does not usually receive a fee, and therefore tends not to regard the other ministry as a client. Nevertheless, it should do so, and this is why the word "client" is retained throughout the Guide.

THE USERS

In many respects the users are the most important party, yet often they are the most neglected. They are the people who must operate and maintain the facilities which have been provided. Although the same organisation may be both client and user, the individuals involved may be different. The planners and administrators of a Ministry of Health may determine how many health centres to build, where they should be located and what facilities should be provided. Although the doctors and nurses operating the centres may belong to the same ministry, they have probably not been involved in any of the early decisions. Conversely, the planners and administrators suffer little inconvenience if the provision of some basic amenity, such as water, has been overlooked; but to the users its provision would be vitally important.

THE DESIGNERS

These are the architects and specialists responsible for translating the client's requirements into reality. In a building project, the architect has a leading role, but he requires support from many other people—

- ☐ draughtsmen to produce working drawings from the architect's sketches;
- ☐ structural engineers in the design of the structure;
- ☐ electrical engineers in the design of power and lighting supplies;
- ☐ civil engineers in the design of access roads, earthworks, water supplies;
- ☐ quantity surveyors in the preparation of estimates and tender documents.

Not every project requires all these people, but very few require only the services of an architect. On the other hand, large and complex projects may require additional specialists. For example, the design of an industrial plant will require considerable expertise in occupational safety and health. However, all those involved at this stage can be classified as designers.

THE EXECUTORS

These are the people who undertake the physical construction, who in many cases will be private or parastatal contractors. Some ministries have their own labour forces, and work carried out in this way is said to be done by "direct labour" (the expression "force account" is also used). Chapter 6 explains this more fully. For simplicity, the word "contractor" is generally used throughout, and is deemed to include all organisations which actually build.

Chapter 6 also explains that the client may place orders for specialised items of equipment directly with a supplier. Examples are medical equipment for a hospital, or desks for classrooms. Such items are as essential to the completion of a project as the buildings themselves. These suppliers are therefore also executors of the project, and must be co-ordinated by the project management team.

PUBLIC AUTHORITIES AND AGENCIES

All buildings must fulfil statutory requirements regarding construction standards and safety. For example, roofs must be able to withstand specified wind loadings, and fire regulations must be observed. Responsibility for ensuring compliance with requirements rests with such bodies as town councils, ministries of planning, water authorities, and so on. The health and safety of people who work or dwell in the building are usually closely safeguarded by legal regulations, and these may have a substantial effect on the shape of the building and the facilities provided.

In some cases a ministry may be simultaneously a client and a statutory authority. A Ministry of Labour which commissions a new labour exchange may also be the authority responsible for ensuring that site conditions of work satisfy employment regulations.

stages and aspects

The major stages of a project, together with the various "aspects" which must be considered during each stage, form the framework of the construction process. These aspects can be divided into four main groups—

functional: general concepts, operational patterns, department and room programmes;

location and site: climate, topography, accessibility, infrastructure, legal formalities;

construction: design principles, technical standards, availability of building materials, building methods, safety of operations;

operational: project administration, cash flow, maintenance needs, operational safety and health.

The examination of each aspect should start during the briefing stage and continue in greater detail during the subsequent stages until each has been dealt with. Each aspect, or group of aspects, will be attended to at different points during the various project stages. To reach a conclusion, it is preferable to follow a formal survey, analysis and recommendations process as shown in figure 6.

When the functional, site, constructional and operational aspects are considered, the degree of detail required should be carefully weighed. Each factor should be examined and its content developed, but only to the

Figure 6. Aspects of a construction project.

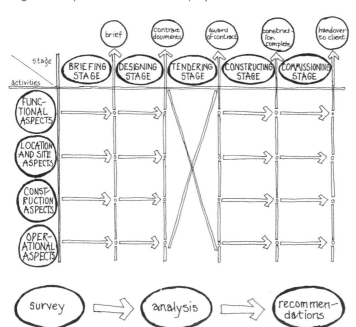

extent needed to fulfil requirements at that particular stage. For example, there is no point in wasting resources on producing a complete room data programme during the briefing stage, since the work during the following stages will significantly influence finishes and installations, and may make much of the early work irrelevant.

Figure 7 shows the sequence of work throughout the different project stages and the people involved.

In many respects each project stage can be considered as an entity, although there is often a degree of overlap between them. The purpose and decisions of each stage should be clearly defined, so that the completion of the stage can be coupled to a commitment or decision by the client.

The work done in the early stages of a project is very important. The relationship between the expenditure and the degree of design influence at the various stages is particularly worthy of note. Once the project has reached the constructing stage, when funds are being spent at a very high rate, it is almost impossible to influence the size and shape of the building. This is illustrated in figure 8. Clearly, the crucial period is when the project brief is being reviewed by the client for final approval. This is the time when substantial savings can be made.

It will be observed that the tendering stage creates a gap in the project process. This is because the call for tenders and their evaluation is frequently handled by an independent Tender Board, over which neither the client nor the project management team has control.

Descriptions of the five stages—their purposes, activities and major participants—are now given in detail.

Figure 7. Participants at each stage of a building project.

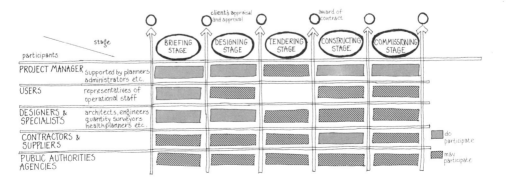

Figure 8. Relationship between design influence and cost.

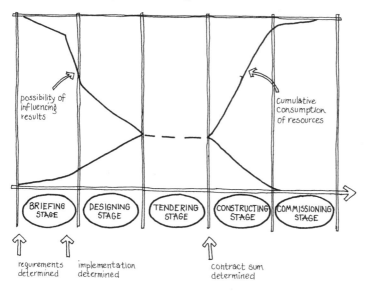

briefing stage

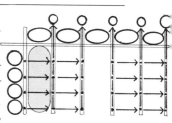

PURPOSE

To enable the client to specify project functions and permissible costs, so that the architects, engineers, quantity surveyors and other members of the design team can correctly interpret his wishes and provide cost estimates.

ACTIVITIES

At this stage many factors of the project are unknown or not clearly defined. Therefore there may be a temptation to postpone the issue of how it is to be managed. As this is probably the most important decision to be reached and one which will contribute to the clarification of other issues, it should be the first task of the client.

The organisation of the management function is described in a later chapter, but a first step is the appointment of a project manager or project co-ordinator who will have a continuing responsibility to the client throughout the whole of the construction process. This person may come from inside or outside the client's own organisation, but during the briefing stage he should—

- ☐ set up a work plan and appoint designers and specialists;
- ☐ consider user requirements, location and site conditions, planning designing, estimated costs, quality requirements;
- ☐ ensure the preparation of—
  - the department data programme;
  - the room data programme;
  - a time-schedule;
  - sketches at scales of 1:1000, 1:1500 or 1:2000, illustrating the layout and principles of the project;
  - cost estimates and implications;
  - a plan for implementation.

Alternative courses of action should be given extra emphasis during this stage.

In some projects the initial user requirements may be unclear, the location uncertain and the cost limit not decided. In such cases it may be helpful to prepare the project brief in steps, first clarifying the major aspects of the project and outlining alternative courses of action and their consequences. Those alternatives that appear most feasible may then be studied further. Such a study should clarify the various aspects of the project in enough detail to enable the project management team to prepare practical recommendations. The details required are given in the briefing stage checklist (Appendix A).

The health, safety and general well-being of the people who will work or dwell in the building should be given careful consideration at the outset. The major influences on design are usually finance, speed, aesthetics, prestige and the available time and knowledge of the designers; occupational health and safety needs are often not considered until the building has been constructed. But diligent attention to these important requirements at the outset will lead to adequate provision at minimum expense. Occupational health and safety are discussed in more detail in Chapter 8.

It is worth stressing that a careful appraisal of user

requirements at the briefing stage makes substantial cost savings easier.

PARTICIPANTS

The main participants at this stage are the client's representatives, the project manager and those responsible for providing technical inputs for the brief.

Depending upon the nature and the complexity of the project, the following should be included:

- [ ] architect;
- [ ] structural, mechanical and electrical engineers;
- [ ] quantity surveyor;
- [ ] specialists such as health or school planners, organisational planners, health and safety officials;
- [ ] user representatives.

See also Appendix A.

designing stage

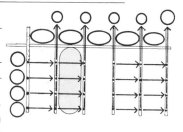

PURPOSE

To complete the project brief and determine the layout, design, methods of construction and estimated costs, in order to obtain the necessary approvals from the client and authorities involved.

To prepare the necessary production information, including working drawings and specifications and to complete all arrangements for obtaining tenders.

ACTIVITIES

Activities at this stage include—

- [ ] developing the project brief to final completion;
- [ ] investigating technical problems;
- [ ] obtaining the client's final approval of the brief;
- [ ] preparing—
    — a scheme design, including cost estimates;

Figure 9. Design-stage activities for building projects.

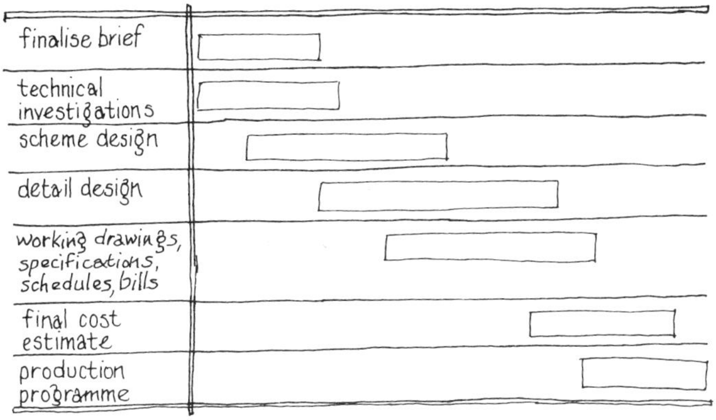

finalise brief	▭
technical investigations	▭
scheme design	▭
detail design	▭
working drawings, specifications, schedules, bills	▭
final cost estimate	▭
production programme	▭

— a detail design;
— working drawings, specifications and schedules;
— bills of quantities;
— a final cost estimate;
— a preliminary production programme, including time-schedule.

These activities are shown in figure 9.

In most projects the designing stage is divided into several substages: outline proposal, scheme design, detail design, and production information. The first two sub-stages are sometimes referred to as "sketch plans", and the latest two substages as "working drawings".

An adequate basis for a realistic cost estimate of the project should be possible from the scheme design sub-stage. Any modification of the project brief after this stage is likely to be expensive.

Regular contact should be maintained between the project management team and the design team, preferably through a series of regular meetings at which progress reports can be considered and any outstanding issues discussed and decided.

PARTICIPANTS

The main participants in the designing stage are the project management team and the design team. Depending on the nature and the complexity of the project, the design team should include the following:

- [ ] project manager;
- [ ] architect;
- [ ] quantity surveyor;
- [ ] structural, mechanical and electrical engineers;
- [ ] specialists such as health or school planners, organisational planners, health and safety officials;
- [ ] user representatives.

See also Appendix A.

tendering stage

PURPOSE

To appoint a building contractor, or a number of contractors, who will undertake the site construction work.

ACTIVITIES

To obtain tenders from contractors for the construction of the building and to award the contract.

During this stage the client enters into firm commitments for most of the project expenditure, and so the procedures and processes are carefully and closely defined.

Government tendering procedures are particularly closely controlled to ensure that national contracts are awarded in an equitable and uniform way. In many cases tendering is the responsibility of a central Tender Board which is independent of either the client or the executing agency, although both may be represented on it. Members of the project management team may be required to provide the basic documentation to the central Tender Board and generally to provide technical assistance.

For simplicity, this Guide describes the management of projects where the construction work is done by a contractor. In some cases a government agency, such as a Ministry of Works, will do the work directly. There may then be no requirement for formal submission to a Tender Board; nevertheless, the executing agency should be required to prepare a detailed cost estimate for budgetary control purposes.

PRE-QUALIFICATION

Some clients will allow any contractor to tender for a contract, provided that he has sufficient financial resources to purchase the tender documents and to give some form of financially backed guarantee that the contract will be accepted if offered. The client then relies on the adequacy of the contract documents and the skills of the project management team to ensure that the building is constructed as required.

To increase the probability that the client *will* get what is required, it is usual to introduce some procedure to ensure that only experienced and competent contractors are allowed to tender. This procedure, known as "pre-qualification", involves an investigation of the potential contractor's financial, managerial and physical resources and of his experience of similar projects, and an assessment of the firm's integrity.

Contractors who satisfy the client's requirements are then placed on a list of approved contractors. Government agencies and other organisations who award many contracts for a wide variety of work may also classify the contractor. For example, a Ministry of Works may have seven categories ranging from Class 7 contractors, who may undertake only very small and simple jobs, to Class 1 contractors, who may undertake any contract regardless of its value or complexity.

CONTRACT DOCUMENTS

It is vital that the contract documents be prepared with extreme care by experienced people, because they will form the major basis on which the project management team exercises control during the construction phase.

The contract itself will be defined in a legal document which describes the duties and responsibilities of the participants who are parties to it. For construction work, standard forms of contract have evolved in most countries, and it is usual for the central Tender Board to require the use of one of these standard forms, with perhaps minor modifications to suit the circumstances of a particular project (see below).

Other contract documents are those necessary to define in detail the building required by the client: drawings, specifications, schedules, bills of quantities, timescale. These documents will have been prepared during the design stage, as described previously.

Standard forms of contract

The use of standard forms of contract is recommended because—

- ☐ their contents will be well known and understood by the parties involved;
- ☐ their wording embodies much experience in resolving difficulties of interpretation and enforcement;
- ☐ they will usually have been tested in law;
- ☐ the preparation of new forms of contract is expensive and time-consuming.

Most standard forms of contract also include the necessary tender documents.

PARTICIPANTS

In the case of government projects the project management team does *not* usually participate, although the Ministry of Works or its local equivalent will be represented on the Tender Board. The project management team may be expected to give technical support by—

- ☐ providing the necessary contract documents;
- ☐ providing a basis for pre-qualification of tenderers;
- ☐ checking that the tenders are arithmetically correct and conform to tender requirements.

Government Tender Boards usually award the contract to the lowest bidder. In the case of private projects the final choice of contractor rests with the client, acting on the advice of the project management team.

See also Appendix A.

constructing stage

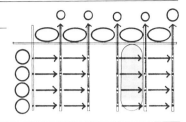

PURPOSE

To construct the building within the agreed limits of cost and time, and to specified quality.

ACTIVITIES

To plan, co-ordinate and control site operations.

Production planning includes the formulation of—

- ☐ time-schedules;
- ☐ site organisation;
- ☐ a manpower plan;
- ☐ a plant and equipment plan;
- ☐ a materials delivery plan.

Site operations include—

- ☐ all temporary and permanent construction works, and the supply of all built-in furniture and equipment;
- ☐ the co-ordination of subcontractors;
- ☐ general supervision.

The constructing stage consists of a number of inter-related activities. The failure of one activity can disrupt the entire production schedule. Therefore, careful production planning is important.

The constructing stage is usually the most expensive, intensive and difficult of the project stages. Later chapters on the organisation of management functions, planning, procurement, control, and health and safety, describe the necessary managerial activities in detail.

PARTICIPANTS

The main participants are the project management team and the contractor.

Normally, direct responsibility rests with the contractor. It is also usual for subcontractors to be directly responsible to the contractor, even if they have been nominated by the client. The project manager and his team must arrange for adequate supervision of the work to ensure adherence to quality standards and to statutory requirements, especially as regards safety and welfare.

See also Appendix A.

commissioning stage

PURPOSE

To ensure that the building has been completed as specified in the contract documents, and that all the facilities work properly.

To provide a record of the actual construction, together with operating instructions.

To train staff in the use of the facilities provided.

RECORDS

During construction, difficulties may arise which result in changes to the original design. Records of these changes will be kept during construction, mainly for financial reasons. These must be brought together to make a complete record of the actual construction.

ACTIVITIES

The activities will be to—

- ☐ prepare "as-built" records;
- ☐ inspect the building thoroughly and have defects remedied;
- ☐ test for watertightness;
- ☐ start up, test and adjust all services;
- ☐ prepare operating instructions and maintenance manuals;
- ☐ train staff.

The test for watertightness is important in areas where rain occurs only at certain seasons. Failure to test at the commissioning stage might result in leaks not being discovered until rainfall occurred—perhaps many months, or even a year, later.

The commissioning stage is the transition period between the construction and the occupation and use of the building. For large and complicated buildings, or groups of buildings, it is not uncommon for the commissioning to be done in several stages.

Commissioning must be planned well in advance so that the recruitment and training of staff and deliveries of furniture and equipment can be co-ordinated with the commissioning schedule.

PARTICIPANTS

The participants are—
- ☐ project management team;
- ☐ operating staff;
- ☐ designers and specialists;
- ☐ building services suppliers staff;
- ☐ the contractor.

See also Appendix A.

A CIVIL ENGINEERING PROJECT 3

It was stated in the preface that this Guide applies to both building and civil engineering projects. It is difficult to define these two categories of construction, but a general understanding of the difference may be gained from figure 1. Broadly, buildings are construction works in which people will work or dwell; civil engineering works are more concerned with controlling the natural environment to provide what is sometimes called the "infrastructure": for example, roads, dams and airports.

It was also stated in the preface that many of the management processes and procedures described in this Guide may be applied to both building and civil engineering projects. It has been assumed—in order to avoid repeating this common material—that the reader has read Chapter 2 on building projects. This chapter will therefore concentrate on civil engineering.

characteristics of civil engineering projects

Because civil engineering work is concerned with changing the natural environment, it is highly susceptible to the unpredictable forces of nature, whereas building works are not so susceptible. For example, the building of a dam may be seriously disrupted by a sudden storm causing a flash flood. But once the foundation of a building is complete, the elements cause minor interference to the erection of the superstructure.

Civil engineering works thus require a more flexible approach as well as an increased contingency allowance for unexpected costs.

Most civil engineering schemes are large, extensive and expensive, obvious examples being road and irrigation developments.

These characteristics often require a high rate of expenditure during the construction stage and a high level of managerial expertise. This may in turn require a high level of investment by the contractor. In developing countries these characteristics have resulted in much civil engineering work being funded by international development agencies, designed by international consultants, and constructed by major international contractors using plant-intensive methods.

CONSTRUCTION TECHNOLOGY

The major civil engineering construction activities are: the excavation, transport and compaction of earth and rock; construction in steel and concrete; and the excavation and laying of pipelines. These activities are generally on a large scale. They are frequently complex and difficult, requiring high levels of engineering skill and expensive modern plant and machinery.

On the other hand, there are many civil engineering projects in developing countries where the technological demands are low. This has led to some successful experiments in the use of low-technology, labour-intensive construction in recent years, particularly in road construction.

This method of civil engineering construction is often more appropriate in developing countries than the system which relies on machinery and complicated processes. Some advantages of labour-intensive methods are—

☐ a reduced level of investment;
☐ the creation of employment;
☐ a reduced need for foreign exchange;
☐ the development of national technical and managerial skills.

In most countries managerial practice in building and civil engineering has developed separately, and therefore many of the differences between the two categories of construction are merely those of historical development.

Thus architects often regard the management of building projects as their prerogative, and engineers take the same attitude towards civil engineering works. Yet many projects involve both disciplines. Fortunately, it is becoming more widely recognised that construction project management is an interdisciplinary art requiring professional services as dictated by the needs of the project. It is this form of project management which is described in this Guide.

participants

The main participants are—
- ☐ the client;
- ☐ the users;
- ☐ the designers;
- ☐ the executors;
- ☐ public authorities and agencies.

The size and scope of civil engineering projects mean that the client is nearly always a government department—often called the "promoter"—or agency, or a major industrial undertaking. Examples are a Ministry of Water Resources or a mining corporation.

THE USERS

The users usually fall into two groups: the operating staff, and those who may be called the "beneficiaries" because they benefit from the scheme although they do not operate it, in the way in which farmers benefit from an irrigation or land improvement scheme. The operators should of course be consulted, but it is perhaps even more important to consult the "beneficiaries". One has only to consider a road being constructed through the rural countryside to realise how much topographical knowledge the local inhabitants can offer to the designers, and how difficult the problems of landownership may become.

THE DESIGNERS

The civil engineer will usually play a leading role in the design team; traditionally he has led it. He will be supported by a number of people, such as—

- ☐ draughtsmen/detailers who make the working drawings;
- ☐ specialists, such as hydrologists, geologists and water treatment chemists;
- ☐ quantity surveyors, or—more usually—"measurement" or "cost" engineers, who are civil engineers who have specialised in this branch of the profession and are involved in the preparation of estimates and tender documents.

THE EXECUTORS

It was explained above that the physical construction will usually be undertaken by a major contractor unless steps are taken at the design stage either—

- ☐ to design the works in such a way that they can be constructed by a large number of smaller contractors working on separate parts; or
- ☐ to design for labour-intensive methods, which will be applied directly by the client or by local contractors.

PUBLIC AUTHORITIES AND AGENCIES

The size, scope and nature of civil engineering work mean that public authorities and agencies play a major, perhaps dominant, part in the project.

stages and aspects

A civil engineering project may be divided into stages similar to those described for a building project, and the aspects to be considered are generally similar also. Thus, the managerial framework given in figure 7 also applies to civil engineering projects, with minor changes, as shown in figure 10. These stages will now be explained in more detail.

Figure 10. Participants at each stage of a civil engineering project.

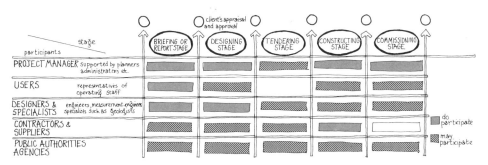

briefing or report stage

PURPOSE

To enable an objective decision to be made on the feasibility of the project, and to examine alternatives.

To specify the client's requirements in detail so that the designers may correctly interpret his wishes and estimate the likely costs.

ACTIVITIES

A civil engineering project should begin with a thorough investigation of its technical and economic feasibility. The characteristics of civil engineering projects described earlier require extensive investigations, such as hydrological and geological surveys, in order that alternative solutions may be evaluated sensibly. Even so, the final outcome of these analyses is often that several, perhaps widely different, solutions are equally attractive. Water supply, for example, may be obtained from boreholes or by building a dam, and it may be difficult to weigh the advantages and disadvantages of each scheme, particularly in the long term.

Investigations and analyses

The following investigations and analyses may be required:

- ☐ non-technical investigations: these include economic and social factors which may define what the nation and its people require of the project;
- ☐ analysis of existing information and plans: for example, reports of previous investigations, maps, charts, official records, seismic records;
- ☐ decisions on standards: all construction work will be subject to standard procedures by which the design methods and materials requirements will be specified; these may be national or international standards, or a mixture of both, and the choice of standards may have a profound effect on the design;
- ☐ general technical investigations: these will include land and geological surveys, and may include specialist investigations by hydrologists, agronomists, or others;
- ☐ detailed site investigation: for example, soil samples and ground water levels from boreholes; access; services;
- ☐ materials: most civil engineering materials (such as earth, rock and concreting materials) are heavy and bulky, and so the proximity of good local supplies may dramatically reduce transport costs;
- ☐ models: models are extensively used in civil engineering to investigate proposed designs in some detail, particularly in schemes involving water.

The Report

Thus the outcome of the first stage of a civil engineering project is the Report, which describes the investigations and analyses that have been done, describes the possible solutions and evaluates them. The Report should also include recommendations to help the client make his choice (see figure 11).

PARTICIPANTS

It is important that a project manager be appointed at the outset in order to ensure that the briefing stage is well managed: the analyses made at this stage, and the solutions and decisions reached, will determine the project's final cost.

Figure 11. The briefing or report stage.

| set objectives and criteria | develop alternative solutions | evaluate alternatives | detailed evaluation | report and recommendation |

In addition, the project manager's ability to manage the remaining stages of the project efficiently will depend in part upon the extent of his knowledge of the fundamental requirements and difficulties of the project.

Depending upon the nature and complexity of the project, the other participants at this stage may include—

- ☐ civil engineers;
- ☐ mechanical, electrical, agricultural and other engineers;
- ☐ non-technical experts such as economists and sociologists;
- ☐ measurement engineers or quantity surveyors;
- ☐ user representatives.

See also Appendix A.

designing stage

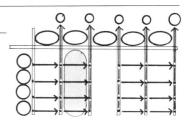

PURPOSE

To design the works in detail, obtain the necessary approvals, estimate costs and prepare the necessary production information.

ACTIVITIES

Activities at this stage will include—
- ☐ deciding which solution to adopt;
- ☐ undertaking further investigations, if necessary.

The following must also be prepared:
- ☐ a scheme design, including cost estimate;
- ☐ a detail design;
- ☐ working drawings, specifications and schedules;
- ☐ bills of quantities;
- ☐ final cost estimates;
- ☐ a preliminary production programme, including time-schedule.

These are shown in figure 12.

In the briefing or report stage, the dimensions and materials to be used in each element of the scheme will have been decided by a combination of approximate analyses and the engineers' experience of similar schemes in the past. This is why consulting engineers with a particular expertise are often employed. The engineering soundness of the chosen outline proposal is then examined in detail, and the design is amended in the light of these analyses.

As the detailed design proceeds, more information becomes available on which an estimate of cost may be based. The accuracy of the estimate will depend upon the

Figure 12. Design-stage activities for civil engineering projects.

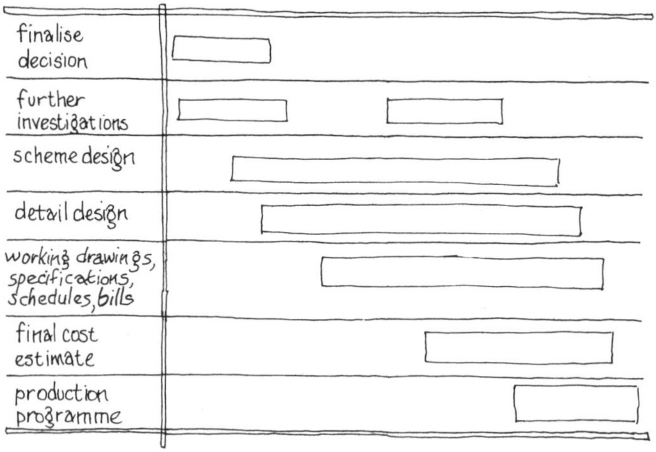

finalise decision	▭
further investigations	▭ ▭
scheme design	▭▭▭
detail design	▭▭▭
working drawings, specifications, schedules, bills	▭▭
final cost estimate	▭▭
production programme	▭

nature of the project and its location. The cost of a small water treatment plant to be built in an urban area may be estimated with reasonable accuracy, because the work will be well understood, and the market prices of most of the required resources will be known. On the other hand, the cost of novel methods of construction to be used in a project in a remote area may be almost impossible to estimate.

When costs are estimated, it is important to make some allowance for the unexpected, because the earlier investigations cannot have covered every possibility. Unless they are done in great detail, and consequently at enormous expense, engineering investigations will be limited in their extent. The project management team should therefore plan the project in such a way that unexpected events do not cause chaos and disruption.

PARTICIPANTS

The main participants are the project management team and the design team. The design team should include the following:

- ☐ project manager;
- ☐ civil engineers;
- ☐ engineers from other disciplines as necessary;
- ☐ specialists;
- ☐ draughtsmen/detailers;
- ☐ quantity surveyors or measurement engineers;
- ☐ user representatives.

See also Appendix A.

tendering stage

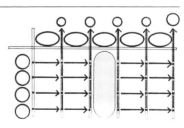

PURPOSE

To appoint a civil engineering contractor, or a number of civil engineering and other contractors, who will undertake the site construction work.

ACTIVITIES

To obtain tenders from contractors for the construction of the works and to award the contract.

PRE-QUALIFICATION

The nature of conventional civil engineering work usually requires some form of pre-qualification of tenders. The pre-qualification procedures will be similar to those described in Chapter 2. The particular objectives of pre-qualification for civil engineering projects are to ensure that those contractors invited to tender have—

- ☐ adequate engineering skills and resources to do the work safely and efficiently;
- ☐ adequate financial and managerial resources to support a project of the scale required;
- ☐ some knowledge of local conditions;
- ☐ integrity.

Where the works have been designed for execution by local contractors, the need for pre-qualification of tenderers is reduced.

See also Appendix A.

constructing stage

PURPOSE

To construct the civil engineering works within the agreed limits of cost and time, and to specified quality.

ACTIVITIES

To plan, co-ordinate and control site operation.

The activities are basically similar to those of "A building project" (see Chapter 2) except that the characteristics of civil engineering projects place more emphasis on the amount and scale of the temporary works. Requirements of health and safety are also more difficult to ensure; paradoxically, these requirements may be more often met in civil engineering than in building because the hazards are more evident to both managers and workers.

PARTICIPANTS

The participants are—
- [ ] the project management team;
- [ ] specialists and specialist engineers involved in the designing stage;
- [ ] the contractor;
- [ ] general and specialist subcontractors.

Most civil engineering projects will require part of the project management team to be resident on site to supervise the construction work in detail.

DOCUMENTS

Standard forms of contract have been developed for general civil engineering contracts, together with special forms of contract and subcontract for specialist engineering work. For example, specific standard forms of contract exist for piling and marine dredging. Standard forms of contract should be used wherever possible, for the reason given in Chapter 2 under "Standard forms of contract".

Government contracts will usually be awarded under procedures controlled by a Tender Board, and many of the private clients who commission civil engineering works will be substantial corporations which adopt similar procedures.

The project management team may be required to provide technical support.

See also Appendix A.

commissioning stage

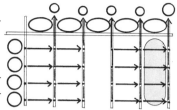

PURPOSE

To ensure that the civil engineering works have been completed as specified in the contract documents, and that all the facilities work properly.

To provide a record of the actual construction, together with operating instructions.

To train staff in the use of the works.

ACTIVITIES

The commissioning stage of a civil engineering project is essentially concerned with testing the engineering performance of the works. Works that are buried as the work proceeds (piling and pipelines, for example) are usually tested before they are covered. Other works can be tested only when complete; water tanks are an example of these.

The commissioning activities for most civil engineering projects will include the following:

- ☐ testing the engineering performance and safety of the whole of the works;
- ☐ searching for non-operational defects, e.g. those that affect only the appearance;
- ☐ remedying all deficiencies;
- ☐ preparing "as-made" drawings and other records;
- ☐ preparing operational instructions and maintenance manuals;
- ☐ training staff;
- ☐ monitoring performance of works against original requirements.

PARTICIPANTS

The participants are—

- ☐ the project management team;
- ☐ the operating staff;

- [ ] designers and specialists;
- [ ] beneficiaries;
- [ ] the contractor.

See also Appendix A.

ORGANISATION OF MANAGEMENT FUNCTIONS 4

objectives

The main objectives of the project management team should include—

- [ ] the production of construction works which satisfy the client's functional requirements;
- [ ] the completion of the project within specified cost limits;
- [ ] the completion of the project within specified time-limits;
- [ ] construction to specified standards;
- [ ] the preservation of the health and safety of the people involved.

Within these main objectives there may be subsidiary objectives for each stage. It is important not only that overall and subsidiary objectives should be clearly stated and accepted by all parties but also that ways of evaluating whether or not objectives have been achieved should be determined at the outset. This is discussed in detail in Chapter 7.

project management team

The need for a project management team is stressed in this Guide, and reference is made in Chapter 1 to the differences between the management of steady-state organisations and that of projects. The team needed will depend on the kind of project to be managed. Many clients, especially those to whom the construction of a building is an unusual activity, underestimate the time, money, skill and effort required. Technical simplicity does not always imply managerial simplicity. For example, a school building programme where the buildings are single storey and of simple block construction may involve the following parties:

- ☐ Ministry of Education—to specify requirements;
- ☐ Ministry of Works—to design the building works and supervise their construction;
- ☐ Ministry of Natural Resources—to supply water;
- ☐ Ministry of Power—to supply electricity, gas or other energy;
- ☐ Municipal Council—to ensure that planning and safety requirements are met;
- ☐ Ministry of Labour—to grant work permits and ensure compliance with labour ordinances;
- ☐ contractors—to put up the buildings and provide access;
- ☐ suppliers—to provide specialist equipment.

There have been health clinics that were unusable because there was no water supply, and bridges that could not be opened because the guard rails had not been ordered.

Project management obviously becomes much more complex as projects increase in size and scope.

TEAM FUNCTIONS

In order to achieve the objectives given above, the management team must exercise the functions of planning, procuring and controlling. These functions will exist through all stages of the project. They are described in detail in later chapters.

TEAM ORGANISATION

Project manager

The first important management decision to be taken by a client is the appointment of a project manager. Few clients understand the need to ensure that their construction projects are properly managed from the start by the right people with defined responsibilities. The work is very different from general administration or production management. When a steady-state activity is being managed, much of the work is routine. The best way to do a particular job has gradually evolved, and so the management task is to facilitate it and to manage the relatively few new circumstances, which may be quite limited in their extent or importance.

In marked contrast, a construction project starts from nothing, builds up to an intense level of activity in the middle of the site construction, and then dies away again to nothing when the buildings or works are taken over by the client. During this time the project manager has to recruit a team of people with the necessary skills to manage the project, a team whose membership will change because of the demands made on it as the project proceeds. This team, and all its supporting offices and facilities, will then be disbanded as the project concludes.

Thus the project manager must be able to deduce the managerial requirements of a new project very quickly and to manage an often intensive and changing series of activities. A specimen job description for a project manager is included as Appendix B, and could be used to formulate terms of reference for his recruitment. When appointing a project manager, the client should consider the following factors:

☐ the qualifications and experience required;
☐ the person or persons to whom he is responsible;
☐ his terms of reference;
☐ the limits of his authority;
☐ his personal qualities, including leadership skills.

As shown in figure 13, the project manager is involved in all stages of the project. The point has been made previously that he should be appointed as early as possible with a view to seeing the project through to completion. Savings resulting from his appointment at a later stage are likely to be offset by increased costs at some critical point caused by his lack of background knowledge of work carried out earlier.

The project manager must thoroughly understand the project. This knowledge should not be limited to its

Figure 13. Steering committee and project management team.

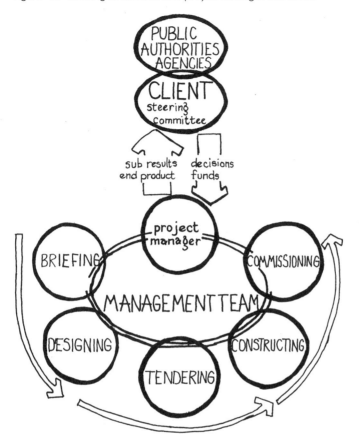

physical features, but should also extend to the client's underlying intentions and policies. He must be able to make logical, impartial and fair decisions so as to achieve the objectives of the project. He must make sure that he has the confidence of his team, so that they, in turn, win the full support and co-operation of their own departments.

No attempt is made in this Guide to specify *who* the project manager should be, since great diversity is a characteristic of building and civil engineering projects. The project manager *might* be: the architect responsible for design; an engineer (from the Ministry of Works); a quantity surveyor; or, especially for a civil engineering project, a qualified civil engineer. In exceptional circumstances, such as the construction of a modern hospital or an extensive civil engineering scheme, the services of professional management consultants may be used. The client must consider each case on its merits and act accordingly.

Steering committee

In government projects particularly, a client ministry or department may have difficulty in fitting a project manager into its organisation. A solution may be to set up an ad hoc steering committee representing the various parties involved. This committee will normally have as chairman a senior official of the client ministry and will include representatives of such other ministries—for example, Works, Finance, Economic Planning—as may be appropriate. The project manager should be a member of the committee, from which he will derive his authority.

The functions of this committee include—

- [ ] determining the terms of reference for the project management team;
- [ ] approving the project management team;
- [ ] monitoring progress of the project;
- [ ] removing obstacles to progress of the project.

Assistants to the project manager

The composition of the team will change as the project progresses through its various stages. The minimum continuous requirement through the life of the project is the project manager, who will require supporting secretarial services. For large and complex projects there may be a need for full-time specialist assistance to the project manager. For example, the construction of a new airport may require a planning engineer to install and run a computerised management information system. A quantity surveyor may be needed to install and run a budgetary control system.

TEAM MEMBERS: BUILDING PROJECTS

Briefing stage

The team would include the following:

- ☐ a client representative;
- ☐ the project manager and supporting staff;
- ☐ specialist assistants to project manager (if any);
- ☐ architects;
- ☐ structural, mechanical and electrical engineers;
- ☐ quantity surveyors;
- ☐ specialists;
- ☐ user representatives—especially if the client is not the final user.

Although most of the team members will not be assigned full time to the project, and may come from different departments or ministries, the project manager should endeavour to instil a team spirit. His ability to do this will depend more on authority derived from a superior grasp of the total project and his human qualities than from paper authority bestowed by, say, a steering committee.

It is an axiom of architecture that form, function and structural integrity must be harmoniously combined in order to produce a wholly satisfactory building. The mechanical and electrical services will also have a significant effect on the form of the building, again demanding a team effort from the designers.

The roles of the team members would be as follows:

- ☐ *client representative*: explanation of the purpose of the buildings, their desired locations, the funds available and any other requirements;
- ☐ *project manager*: general co-ordination, preparation of a work plan, securing client's approval of brief;
- ☐ *architect*: assessment of client's functional requirements, design for pleasing aesthetic appearance (inside and out), constructional soundness of building materials and methods (although the structural engineering analysis will usually be done by a structural engineer);
- ☐ *structural engineer*: working closely with the architect to determine the most satisfactory combination of functional requirements and structural form;
- ☐ *mechanical and electrical engineers*: principally concerned with maintaining a satisfactory environment *within* the building (heating, cooling, ventilation, light-

ing) and with the equipment required for the building's function: for example, welding equipment in a workshop or lifts in an office building;
- ☐ *quantity surveyor:* estimates of cost based on previous cost data from similar buildings, assessment of extra costs due to special features, formulation of a method of cost control;
- ☐ *specialists:* investigation and specification of special requirements such as provision of X-ray equipment in a hospital, or acoustics in a lecture hall;
- ☐ *user representative:* specification of user requirements in detail, supply of background information.

Designing stage

The team would include the following:
- ☐ the project manager and supporting staff;
- ☐ specialist assistants to project manager (if any);
- ☐ architects;
- ☐ quantity surveyors;
- ☐ structural, mechanical and electrical engineers;
- ☐ specialists such as health or school planners;
- ☐ user representatives;
- ☐ (contractor).

At this stage, the architects and the quantity surveyors would normally be full-time members of the team. The engineers may or may not be, depending on the size and complexity of the project. The specialists and user representatives would be part-time members. For the contractor to be a member of the team implies that he has been selected. There are strong arguments for being able to draw on a contractor's experience at this stage. He could advise on construction methods so as to influence the design and thus effect considerable overall cost savings. This would mean using either the "early selection" or the "design and construct" approaches to appointing contractors described in Chapter 6 below. This is a very controversial issue.

The roles of the team members would be as follows:

- ☐ *project manager:* co-ordination of the design stage activities shown in figure 9 (this is very important, as it

is often wrongly assumed that design work cannot be planned), elaboration of work plan, securing client's agreement to final drawings and estimates;
- ☐ *architect*: translation of client brief into, first, outline proposals, and second, working drawings;
- ☐ *quantity surveyor*: costing of outline proposals, preparation of bills of quantities and tender documents, costing of working drawings;
- ☐ *structural engineer*: design of structure and preparation of working drawings based on architect's plans;
- ☐ *electrical engineer*: design and preparation of working drawings for electrical power and distribution systems;
- ☐ *mechanical engineer*: design and preparation of working drawings for heating, ventilating, air conditioning, lifts and other mechanical services;
- ☐ *specialists*: information regarding special requirements for such items as occupational health and safety, medical equipment, teaching aids and workshop equipment;
- ☐ *user representatives*: information regarding such matters as room layouts, access, environmental conditions;
- ☐ *contractor*: advice on construction materials and methods.

Tendering stage

On government projects using the standard approach, the project management team would not normally be involved in this stage. It is usual for an independent Tender Board to invite and award tenders. It would be the responsibility of the project management team to supply the Tender Board with the necessary contract documents, this being the final activity of the designing stage. *After award of the contract, the quantity surveyors would check the documentation for arithmetical errors before formal notification to the successful contractor.*

In the case of a non-government contract, the project management team would include the following:

- ☐ the project manager;
- ☐ quantity surveyors;
- ☐ (contractor).

Constructing stage

The team would include the following:
- [ ] the project manager and supporting staff;
- [ ] specialist assistants to project manager (if any);
- [ ] clerks of works;
- [ ] quantity surveyors;
- [ ] the contractor;
- [ ] (resident architects).

At this stage, the efforts of the project management team should be directed towards construction within time and cost limits, to standards all as specified in the contract documents, and within a basic humanitarian framework of health and safety. The contractor plays a major role, since it is he who is legally responsible for the building work. Nevertheless, the project manager should endeavour to create a team spirit so that all members are working to a common goal. Dissension between the members will not be in the interests of the client.

The roles of the team members would be as follows:

- [ ] *project manager:* agreement with the contractor on site organisation and production schedules, monitoring of progress, co-ordination with designers and statutory authorities, co-ordination with quantity surveyor in preparation of cash flow charts and valuations, information to client regarding progress and costs;
- [ ] *clerk of works:* supervision of contractor to ensure achievement of quality standards in accordance with specifications, technical advice to contractor on interpretation of drawings and specifications, liaison with architect on design matters, recording of variations from original design;
- [ ] *quantity surveyor:* periodic (usually monthly) valuation of work to date, preparation of certificates for interim payments to the contractor, preparation of cash flow statements for client, measurement and valuation of variations, preparation of final account;
- [ ] *contractor:* construction of the building (including all temporary and permanent works), co-ordination and payment of subcontractors, preparation of pro-

duction schedules to the satisfaction of the project manager, general supervision, provision of information as required under the terms of the contract.

The internal management and organisation of a contracting company are separate subjects which are not dealt with in this Guide.

Commissioning stage

The team would include—
- [ ] the client or his representative;
- [ ] the project manager and supporting staff;
- [ ] specialist assistants to project manager (if any);
- [ ] users;
- [ ] designers and specialists;
- [ ] the contractor and subcontractors.

This is an important but frequently neglected stage. The client must be satisfied that he has received what he has paid for. The users must understand how to operate and maintain the facilities. In a complex building, this may require special training; but even for a simple building, such as a rural health centre, the users must know what the various rooms are intended for and how, for example, the water supply works. The contractor must obtain a formal handing-over certificate because he will be responsible for making good any defects found during a specific period (usually six months or a year) from this date. A sum of money will be withheld from him until the end of this period to ensure his performance.

The roles of the team members would be as follows:
- [ ] *client*: liaison with project manager to ensure that the building has been satisfactorily completed, contractor payments;
- [ ] *project manager*: co-ordination of commissioning activities;
- [ ] *users*: training of operating and maintenance staff, acceptance of buildings and installations;
- [ ] *designers and specialists*: checking that buildings and installations conform with contract requirements, in-

forming users of special features regarding operating and maintenance;
- ☐ *contractors and subcontractors:* remedying defects, submitting final financial statements and receiving payments.

TEAM MEMBERS: CIVIL ENGINEERING PROJECTS

A project management team for a civil engineering project functions in very much the same way as that for a building project. This section of the Guide describes the main differences, based on the following general list of team members:

- ☐ client representatives;
- ☐ the project manager and supporting staff;
- ☐ specialist assistants to project manager (if any);
- ☐ civil engineers;
- ☐ other engineers;
- ☐ experts from non-engineering disciplines;
- ☐ user representatives.

Project manager and supporting staff

The civil engineering client may be very knowledgeable about the functional requirements of the project. For example, the client for a water supply scheme will usually be a government ministry employing enough qualified and experienced people in all relevant disciplines to manage the project entirely "in-house". Thus the client's representatives may play a large part in the management of the project; indeed, it is quite common for the project manager to be a member of the client's permanent staff.

Civil engineers

Building project managers may be chosen principally for their managerial abilities. The nature of civil engineering projects usually demands a technically competent civil engineer as project manager. The technical demands of civil engineering projects will also be reflected in the predominance of civil engineers in the project team.

Figure 14. Progress of a civil engineering project based on a possible water supply scheme.

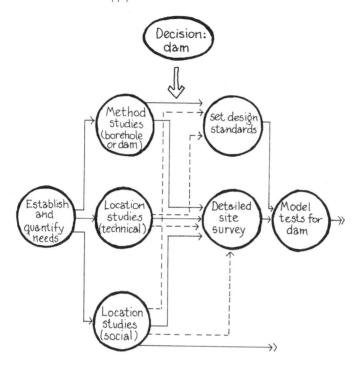

Other engineers and non-engineering experts

The investigations and analyses that may be required for a civil engineering project were described in Chapter 3. Consideration of these will give some indication of the co-ordination difficulties that a project manager and his team may face in a large and complex project. As an example, figure 14 illustrates the investigations needed to decide on a water supply scheme. Much of the work to be done is technically difficult; it may be time consuming; subsequent changes may be delayed until results of the investigations and analyses have been carefully studied.

User representatives

It may also be necessary for the project management team to consult with user representatives, and this may result in unpredictable and time-consuming delays.

the reality of management

The processes and procedures described in this Guide form a basis for good management of construction projects, but they cannot in themselves manage a project. Projects are managed by people who have to make decisions and enforce procedures that affect other people. Project management must be seen as a dynamic, difficult and often abrasive art, based on well-proven principles but not solely devoted to their slavish or rigorous application.

PLANNING 5

Planning, as shown in figure 15, is the spine of the whole project, and must be based on clearly defined objectives. With proper planning, adequate resources are available at the right moment, adequate time is allowed for each stage in the process, and all the various component activities start at the appropriate times.

Planning should include—

- ☐ forecasts of resource requirements of people, materials and equipment; analyses for their most efficient use;
- ☐ forecasts of financial requirements;
- ☐ provision of "milestones" against which progress can be measured.

Figure 15. Planning throughout the project.

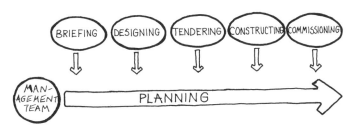

participants

The project manager will be responsible for the planning process during all stages of the project. He and his team will be assisted during briefing, designing, tendering, constructing and commissioning by others, as described in the earlier chapters.

principles

Planning techniques range from simple bar charts to computerised network analysis. Some reference is made to these later, but all the techniques are based on certain important principles—

- ☐ the plan should provide information in a readily understood form, however complex the situation it describes;
- ☐ the plan should be realistic. There is no point, for example, in planning a building to be completed in six months if the delivery period for cement is five months;
- ☐ the plan should be flexible. Circumstances will almost inevitably change during the constructing stages. It should be possible to alter certain elements without disrupting the entire plan;
- ☐ the plan should serve as a basis for progress monitoring and control;
- ☐ the plan should be comprehensive. It should cover all the stages from briefing to commissioning. It is a common misconception that planning is necessary only for actual construction. Even on a small project, the time between the decision to build and the taking-over of the completed works is often two or three years. Out of this period, only nine months or a year may have been spent on physical construction. On a large and complex project the proportion of total time spent on building may be as low as 20 or 25 per cent. The rest of the time will have gone on planning permissions, compliance with statutory requirements, financial authorisation, design tendering, and so on. It is therefore essential to plan the total project period.

techniques

The main techniques used for planning building projects are—

- ☐ bar charts, sometimes known as Gantt charts;

□ network analysis, sometimes known as critical path methods (CPM) or programme evaluation and review technique (PERT). The diagrams may be "activity-on-arrow" or "precedence", and follow the same principles.

More detailed descriptions of these techniques are given in Chapter 10, but an outline is given below.

BAR CHARTS

Most projects, however complex, start by being depicted on a bar chart. Even when a more sophisticated technique is necessary for detailed planning, the results are often shown in bar-chart form. The principles are very simple, and were illustrated in figures 9 and 12.

The following steps are taken:

□ a list of project activities is prepared;
□ the time and resources needed for each activity are estimated;
□ each activity is represented by a horizontal bar drawn to a time-scale;
□ activities are plotted on a chart with a horizontal time-scale. It is then possible to see when they are *planned* to start and end.

CRITICAL PATH ANALYSIS
("ACTIVITY-ON-ARROW" METHOD)

As with bar charts, the first step is the preparation of a list of project activities. However, an important difference from bar charts is that estimates of the time and resources needed for each activity are not usually made at this stage. Instead, each activity is represented by an arrow, *but not to a time-scale*. The tail of the arrow represents the start of an activity and the head its end. The arrows are then arranged to depict the logical sequences of activities, thus producing a network as shown in figure 16. It is only at this stage that estimates of duration and resources for each activity are added. It then becomes possible to calculate the *shortest time* needed to complete the project, and the sequence of activity necessary to achieve this. This is

Figure 16. A simple network planning diagram.

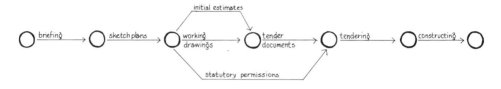

known as the *critical path*. A fuller explanation of this technique is given in Chapter 10.

OTHER PLANNING TECHNIQUES

Time-chainage charts are used extensively for the planning and control of linear construction works such as roads and tunnels. This technique is also described in Chapter 10.

There are other techniques which can be employed, but many are adaptations or refinements of those mentioned above. None has gained wide acceptance, however, and so they are not described in this Guide. Project managers in developing countries will be more effective if they concentrate on understanding the fundamental objectives of the project, and apply simple techniques for its planning and control rather than trying to find "magic" techniques to solve their problems.

activities

The planning of project activities should cover the following:
- ☐ time;
- ☐ briefing and design capacity;
- ☐ constructing and commissioning capacity;
- ☐ supply of equipment and materials;
- ☐ allocation of funds and estimation of costs;
- ☐ staffing and services.

TIME PLAN

The most important task in the planning process is the preparation of a realistic time-schedule. A basic time-schedule should be worked out at a very early stage and should serve as a framework within which all key activities can be indicated.

A series of time-schedules will be prepared during the different stages of the project. During the initial stages the time-schedules will be less detailed but, as the project proceeds, more and more information will be available and details in the time-schedules can be refined and complemented.

Some of the activities, such as financing, approvals by the client and the procurement of furniture, plant and equipment, are not always covered in the above schedules. In such cases it is the responsibility of the project management team to make sure that these activities are given proper attention. The project management team should maintain a time-schedule clearly showing the activities that the team is controlling.

Briefing stage

The interval between the decision to build and the actual taking-over of a completed project is seldom less than two to three years, even for small projects. Therefore, during the briefing stage it is necessary to prepare a time-schedule which, apart from indicating the major project stages, provides for such activities as obtaining planning permission, financing and the procurement of plant, equipment and furniture. The time-schedule should also allow adequate time for any appraisals and approvals between each project stage which are required by the project management team, the steering committee and the client. Unfortunately, such provisions are not often made, and this results in delays and shortages of funds.

Designing stage

During the first part of the designing stage, the designing team should prepare, in collaboration with the project management team, a time-schedule covering in detail the activities up to and including the proposed tender action.

This time-schedule should also indicate the activities to be undertaken by the different designers and specialists in the designing team, and the activities required from the project management team, clearly showing where activities are dependent upon each other. Working drawings and tender documents are prepared during the second part of the designing stage. The time planning during this stage should also cover the tendering activities and the constructing stage.

Tendering stage

The time-schedule should include such activities as calls for tenders, the evaluation of tenders, and the award of the contract. It should also include the overall plans of construction and commissioning.

Constructing stage

During the constructing stage the contractor should prepare a detailed time-schedule based on the time-limits stated in the contract. This time-schedule should indicate how the construction will proceed, including such activities as the installation of plant and equipment, and the advance procurement of any materials.

The project manager must ensure that any materials or equipment that have not been included in the construction contract are obtained in time. He should also ensure that the provision of facilities such as roads, water and electricity is considered during the preparation of the time-plan at the constructing stage.

Commissioning stage

In the commissioning stage the project manager will be responsible for planning activities which take place after the work has been completed and handed over.

BRIEFING AND DESIGNING CAPACITY

One of the first things the project manager should consider is the available briefing and design capacity. A public client must consider the available capacity and competence within his own organisation before employing any external resources. The selection and employment of designers and specialists are discussed in the next chapter.

CONSTRUCTING AND COMMISSIONING CAPACITY

The size and organisation of the construction industry vary from one country to another. Smaller projects are often constructed on a self-help basis or through direct labour schemes. Larger projects are normally constructed using more permanent agencies such as construction units or contractors. Whichever method is selected, it is the responsibility of the project manager to consider the capacity of the selected agency, and to take this into account when the basic time-schedule is prepared. It should be noted that the choice of construction method may have a significant impact on the way in which the production documents are eventually prepared.

A project is not complete until it is operational. The project manager must therefore also consider requirements for such activities as the testing and commissioning of water supplies, mechanical and electrical services, communication facilities, and so on.

SUPPLY OF EQUIPMENT AND MATERIALS

Many projects are not completed on time owing to a lack of, or delay in the delivery of, vital materials. In many cases these delays could have been avoided if the procurement of material supplies had been properly planned. During the designing stage the project manager should go through the list of major materials and items of equipment in the project, and check whether these will be available. If there is a likelihood that any may be difficult to obtain, the project manager should act to avoid possible delays. When shortages of key materials cannot be avoided, the project manager should ensure that the schedules reflect this.

FUNDS AND COSTS

Whether funds for the project are being provided by an aid donor, private capital, a bank or a government ministry, it is necessary to draw up a total budget showing what funds are needed and when. Funds must be made available not only for actual construction but also for the payment of the managers, designers and specialists, and for plant, water, electricity and equipment. Figure 17 shows these requirements. The time-schedule will provide the first indication of when funds should be made available, and the designing team will be able to assist with an assessment of how much money should be made available at the different project stages. Project funds are further discussed in the next chapter.

Figure 17. Requirements for funds.

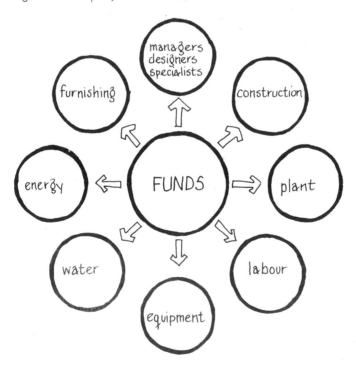

STAFFING AND SERVICES

Planning for use of the building should start during the briefing stage. A list of the staff, staff quarters, infrastructure and services needed for the effective use of the completed works should be prepared, and this should form part of the briefing documents. Using this list, the management team should make plans to ensure that adequate staff and services will be available. If it is found that it will not be possible to obtain these by the time the project is completed, alternative measures may have to be considered. The completion date for the whole project could, for example, be put back; or it might be possible to complete the project in phases.

PROCUREMENT 6

objectives

A construction project requires technical and managerial know-how, manpower, materials, plant and equipment. If, as is usually the case, the client's own organisation cannot provide these resources, they have to be procured (i.e. bought). Although the acquisition of all the resources needed is the ultimate responsibility of the client, he may employ agents for this purpose. The objective of procurement is to ensure that the resources for the project are acquired in the most effective way. Consideration must be given to quality and completion time as well as cost. Figure 18 shows the resources to be procured.

However, decisions on procurement are not based solely on economy, efficiency and quality. Some clients, particularly public clients, consider other objectives, such as the development of local contractors, the use of local

Figure 18. Procurement: the acquisition of project resources.

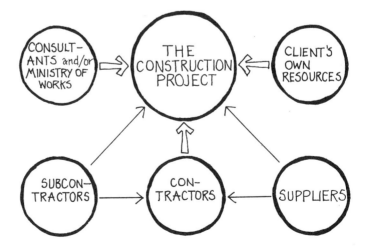

materials or the provision of training for their own staff. This may increase the cost of the project but in the long run may help the development of the local construction industry and the national economy. If objectives other than those related to the efficiency of the project are to be considered, the client should provide the project manager with clear guide-lines.

participants

It is a requirement of many government contracts that client ministries wishing to build should first approach the Ministry of Works or its equivalent. If the Ministry of Works is unable to provide the architects, engineers and other specialists from its own resources, it may procure them from private consultants. Either way, the Ministry of Works is in effect the consultant of the client ministry. Alternatively, the Ministry of Works may authorise the client ministry to employ its own consultants directly.

It is the responsibility of the project manager to ensure that all resources are procured as needed. He may be required to advise the client on the appointment of consultants, and he may also assist the Tender Board in the evaluation of tenders from contractors and suppliers.

methods

The resources required for a construction project fall into two major categories. First, technical and managerial expertise are needed; and second, construction labour, plant, equipment and materials must be provided.

These resources may be acquired in various ways which are described below.

THE STANDARD APPROACH

The standard approach to a building project (see figure 19) is described earlier in the Guide. In this process the client procures external resources on just two occasions—once at the appointment of consultants, and once

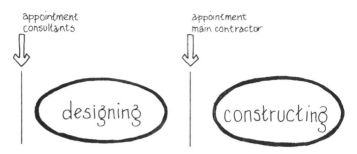

Figure 19. The standard approach.

at the tender action for appointment of a main contractor. Normally separate contracts are negotiated with each consultant, but the co-ordinating function is entrusted to one of them. The main contractor is responsible for the implementation of all construction work, including the work of subcontractors.

The standard approach is characterised by the preparation of complete working drawings and production information before seeking tenders. This facilitates the tender procedure and administration of the contract. This approach has proved generally successful and is used on most government projects. However, the need to shorten the project implementation time has encouraged the development of several other approaches, some of which are now described.

THE EARLY SELECTION APPROACH

This enables site operations to be started earlier than traditionally possible, through the overlapping of the designing and constructing stages, as shown in figure 20. One possibility is by early selection of the main contractor. However, as production information is not complete when the main contractor is appointed, it is difficult to obtain a fixed price for the works. Alternative forms of payment are more complicated and will increase the risk of disagreement between the client and the main contractor. This approach may be useful for a project consisting of a series of similar subprojects.

Figure 20. The early selection approach.

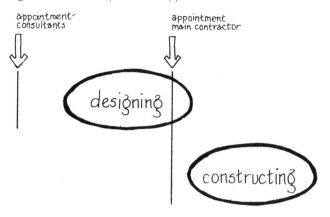

THE DESIGN-AND-CONSTRUCT APPROACH

In this system, sometimes called a "package deal" or "turnkey" project, the contractor is responsible for both design and construction, and thus better co-ordination is possible (see figure 21). This approach may, however, be difficult to combine with a competitive tendering procedure, and the client has to rely heavily on the integrity and competence of the contractor.

Figure 21. The design-and-construct approach.

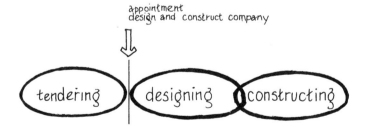

THE DIVIDED CONTRACT APPROACH

In the divided contract approach (see figure 22) a separate contractor is appointed for the earthworks so that site operations can be started earlier than would normally be possible. The number of contracts may be increased;

Figure 22. The divided contract approach.

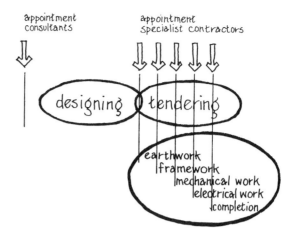

and procurement will then be phased between several different contracts, such as for the building frame or installations. However, by using the divided contract approach the project manager will face greater problems of co-ordination than with a standard approach, since responsibility on the construction site will be divided between several contractors.

THE DIRECT LABOUR APPROACH

Smaller projects in developing countries are often executed by means of direct labour. This approach is particularly suitable for technically simple projects located in remote areas. Since there is no main contractor, the required production information and "bill of quantities" may be simplified. This approach is normally unsuitable for larger projects, as construction capacity is often limited by a lack of plant and skilled staff.

Certain clients regularly involved in construction projects have found it useful to establish their own permanent construction units. If these are well equipped and staffed, they may also undertake larger or more complex projects.

appointing consultants

Many client ministries do not possess the expertise necessary for the direct supervision of a construction project. For the technical expertise (architects, engineers, quantity surveyors), it will often be a requirement that the ministry should first seek help from the Ministry of Works. If that Ministry is unable to carry out the necessary design work, it may authorise the client ministry to engage consultants. The acquisition of managerial expertise by the client ministry may be by the appointment of an existing staff member with appropriate knowledge and experience, by recruitment from outside the ministry, or by the appointment of project management consultants. Once selected, the project manager should be involved in the procurement of design resources.

If the Ministry of Works is able to provide the necessary design services, its staff must identify with the aims and objectives of the client ministry. The role of the project manager in ensuring that a proper brief is provided to the Ministry of Works, and in planning the whole project, remains the same.

The selection of consultants is an important and difficult task. Experience from earlier projects is a decisive factor in this choice, and a contract may often be concluded after negotiations with only one firm. One recommended method of appointing consultants is to ask those being considered to provide the following information, thereby allowing comparisons to be made between them (see also figure 23):

- ☐ the capacity of the firm, which should be large enough to undertake the projects;
- ☐ earlier experience from a similar kind of project, and references;
- ☐ a proposal of the methods which would be used to carry through the project, including methods for co-ordination and control;
- ☐ the proposed project organisation, including the names and qualifications of key staff.

Using this comparative information, and after checking the consultants' references, the client should be able to make a suitable choice. For projects for which a team of

Figure 23. The selection of consultants.

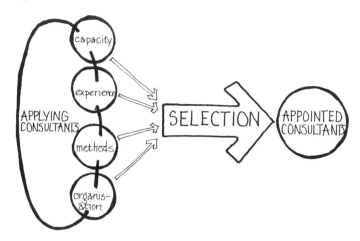

consultants is to be appointed, it is important to obtain a combination of firms which can show that they have worked together successfully on previous occasions. The entire project management team could be appointed at the same time or, alternatively, one consultant could be appointed to assist the client to select and appoint the remaining members.

CONTRACTS BETWEEN CLIENT AND CONSULTANT

Even with a confident relationship between the client and his consultant, it is advisable to establish a formal contract between them. This contract can normally be based on one of the standard forms of agreement prepared by the various professional institutes, e.g. the IGRA, or *International General Rules for Agreement between client and consulting engineer* (1976), prepared by the International Federation of Consulting Engineers (known as FIDIC, after its French acronym).

PAYMENT OF CONSULTANTS

Normally payment is based on scales of fees, which are determined as percentages of the final construction cost, or part of it. The actual percentage varies with the services provided and the type of project. The methods of calculating the fees are defined by the various professional institutes. However, a disadvantage of fees related to construction cost is that a consultant who puts extra effort into the design work in order to find a less expensive solution will receive a smaller fee.

appointing contractors

The recommended procedure for the appointment of contractors is through competitive tendering by a selected number of pre-qualified tenderers. Sometimes public organisations use a register of contractors who are considered suitable for various types and sizes of project. It is also possible to select tenderers from those contractors who respond to an advertisement inviting tenders for a particular project. In any event, pre-qualification is desirable to eliminate unacceptable contractors and to encourage serious prices from others. The tendering process is shown in figure 24.

It is advisable to select prospective tenderers well before the tender documents are to be sent out. It is also important to give sufficient time for the preparation of tenders. The time needed depends on the size and complexity of the job, but generally about four weeks will be required.

Figure 24. The tendering process.

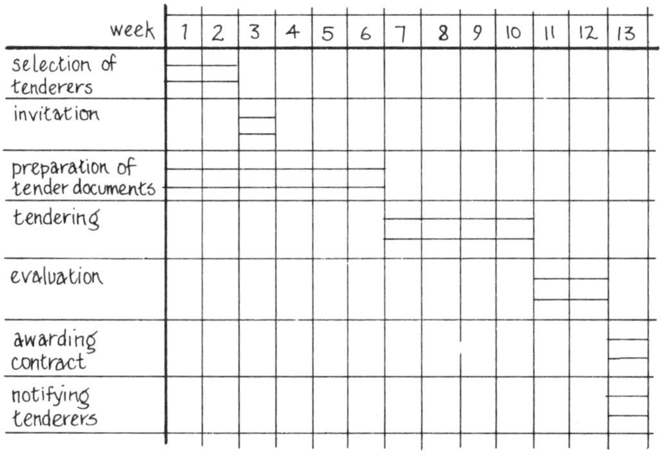

week	1	2	3	4	5	6	7	8	9	10	11	12	13
selection of tenderers	═	═											
invitation			─										
preparation of tender documents			─	─	─	─							
tendering						─	─	─	─				
evaluation										─	─		
awarding contract												═	
notifying tenderers												═	═

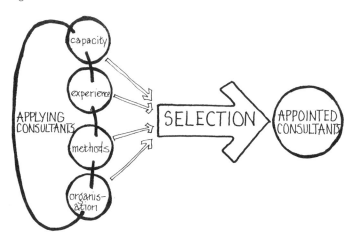

Figure 23. The selection of consultants.

consultants is to be appointed, it is important to obtain a combination of firms which can show that they have worked together successfully on previous occasions. The entire project management team could be appointed at the same time or, alternatively, one consultant could be appointed to assist the client to select and appoint the remaining members.

CONTRACTS BETWEEN CLIENT AND CONSULTANT

Even with a confident relationship between the client and his consultant, it is advisable to establish a formal contract between them. This contract can normally be based on one of the standard forms of agreement prepared by the various professional institutes, e.g. the IGRA, or *International General Rules for Agreement between client and consulting engineer* (1976), prepared by the International Federation of Consulting Engineers (known as FIDIC, after its French acronym).

PAYMENT OF CONSULTANTS

Normally payment is based on scales of fees, which are determined as percentages of the final construction cost, or part of it. The actual percentage varies with the services provided and the type of project. The methods of calculating the fees are defined by the various professional institutes. However, a disadvantage of fees related to construction cost is that a consultant who puts extra effort into the design work in order to find a less expensive solution will receive a smaller fee.

appointing contractors

The recommended procedure for the appointment of contractors is through competitive tendering by a selected number of pre-qualified tenderers. Sometimes public organisations use a register of contractors who are considered suitable for various types and sizes of project. It is also possible to select tenderers from those contractors who respond to an advertisement inviting tenders for a particular project. In any event, pre-qualification is desirable to eliminate unacceptable contractors and to encourage serious prices from others. The tendering process is shown in figure 24.

It is advisable to select prospective tenderers well before the tender documents are to be sent out. It is also important to give sufficient time for the preparation of tenders. The time needed depends on the size and complexity of the job, but generally about four weeks will be required.

Figure 24. The tendering process.

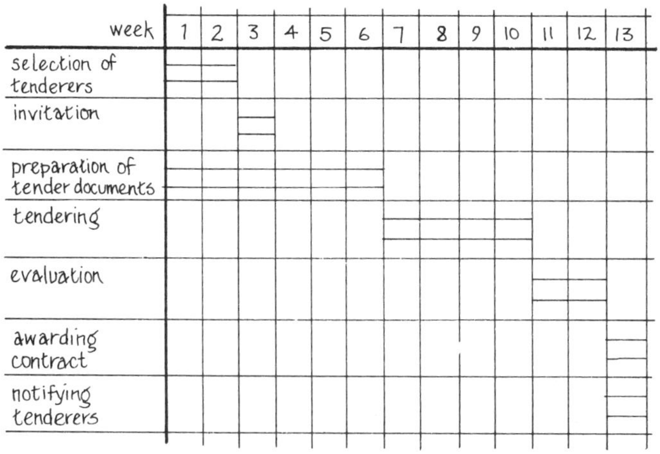

week	1	2	3	4	5	6	7	8	9	10	11	12	13
selection of tenderers	██												
invitation			▬										
preparation of tender documents			▬▬▬▬										
tendering							▬▬▬▬						
evaluation											▬		
awarding contract												▬	
notifying tenderers													▬

Competitive tendering is governed by strict rules, the general aim of which is to guarantee fair competition and unbiased tender evaluation. In many countries, tenders for government projects are usually submitted to and evaluated by a Tender Board which is independent of client and executing ministries, although the Ministry of Works is normally represented on the Board.

When all the tenders come from a selected number of well-known contractors, it is normal to accept the lowest tender. However, an evaluation should also be made of other information such as—

- percentage additions to prime-cost items;
- the amount of liquidated damages per day or other specified period;
- cash advances required;
- time for completion;
- quality of contractor's past workmanship.

If quantities of work done are used as the basis for calculating the price, the project management team should examine the priced "bills of quantities" of the contractor who has submitted the apparently most favourable tender in order to determine any errors or abnormal prices.

Arithmetical mistakes should be corrected and the tender modified on this basis. If, however, there is an obvious mistake in the unit price for an item, the tenderer should be given the opportunity to withdraw the tender, but not to correct the price.

NEGOTIATED CONTRACTS

When it is difficult to specify the scope of work, it may be necessary to negotiate a contract without competitive tendering. Since this implies a higher price, and does not guarantee the same objectivity, it is not so often used for contractor appointment. In certain cases, however, it may offer advantages: for example, when a similar project has recently been successfully completed. Time and effort are then saved if a new contract is negotiated directly with the same contractor on the basis of the earlier one. A negotiated contract may also offer advantages if a contractor in whom the client has confidence can offer the special competence required to undertake a project.

CONTRACTS BETWEEN CLIENT AND CONTRACTOR

In many countries different standard forms of contract are available, depending on the type of project or the form of payment. These can include—

- ☐ standard forms of building contract with or without quantities included;
- ☐ standard forms of contract for civil engineering works;
- ☐ cost reimbursement contracts;
- ☐ contracts for electrical and mechanical works.

International conditions of contract for civil engineering works, and for electrical and mechanical works, have also been established by FIDIC and other international organisations and contractors.

The contract will normally include the following items:

- ☐ an agreement between the client and the contractor;
- ☐ the tender;
- ☐ standard conditions of contract;
- ☐ particular conditions of contract;
- ☐ bills of quantities;
- ☐ a schedule of prices;
- ☐ drawings and specifications.

PAYMENT OF CONTRACTORS

Payments to contractors are normally made on a fixed-price basis in one or other of the following forms: *either* a schedule of prices giving unit prices for most items in the bill of quantities; *or* a lump sum for the complete works as defined by drawings and specifications.

When the payment is based on a schedule of prices, the total price is determined by the actual quantities, measured on the site, multiplied by the unit prices in the contract. During the construction works interim payments are normally made monthly according to valuations of the work completed. This procedure is described in the next chapter. The use of a "schedule of prices" makes interim valuations and the pricing of variations easier and more accurate.

In the event of early contractor selection, a complete bill of quantities cannot be prepared since the design of the project is not finalised. An approximate bill of quantities must be used instead. The final price is calculated from the actual quantities and unit prices included in the bill. When the prices are incomplete or not applicable, new prices have to be established by a rather complicated process of negotiation.

For smaller projects, or for projects where it is difficult to break down the work into units, the price is often in the form of a lump sum. This form of payment is not advisable when the quantities of work are uncertain, because, for example, of the lack of information on soil conditions. In such cases either the client or the contractor has to take the risk of increased costs. With a "lump sum" payment the contractor takes the risk, and to cover himself against this possibility he increases his tender price accordingly.

A "fixed price" contract is not suitable for all kinds of project. It may, for example, be necessary to appoint a contractor before the scope of the work can be defined satisfactorily. The only choice will then be a "cost reimbursement" contract, where the contractor is paid for his verified costs plus a predetermined additional fee. This additional fee normally covers a defined part of the contractor's overheads, and his profit. The fee can be either in the form of a lump sum or as a percentage of the verified cost.

The danger of uncontrolled cost increases when a cost reimbursement form of payment is used may be reduced with a form of target contract. The additional fee is then determined in such a way that the client and the contractor share the difference between the actual cost and the agreed estimated cost.

In times of high inflation, or for projects with a construction time exceeding one year, tenderers normally require some form of compensation for price escalations. If there is no fluctuation clause in the contract, the tenderer will increase his initial tender price. If an official system for indexing construction costs exists, it is often advisable to accept a fluctuation clause in the contract. If no such indexing system exists, it will be necessary to calculate actual increases in material costs, wages and overheads, which is often a contentious and time-consuming activity.

appointing subcontractors

When the main contractor is appointed by the client according to the standard approach described earlier, he assumes total responsibility for the construction works. As he usually cannot undertake the complete work with his own labour, he will need subcontractors. The installation of plumbing and electrical services are examples of work often done by subcontractors.

There are two main procedures for appointing subcontractors—first, for those appointed directly by the main contractor without the involvement of the client; and second, for those nominated by the client before or after the main contractor is selected.

In both cases the main contractor is responsible for the ordering, and for the satisfactory completion, of the work by the subcontractor.

Nomination of a subcontractor gives the client more control over the choice of materials and components, and over the quality of the work. Certain design work may also be undertaken by the subcontractor. Time can thereby be saved where certain components take a long time to make.

A frequent disadvantage with the nomination of subcontractors is lack of competition. This may be overcome if subcontractors are nominated through competitive tendering. Nomination often involves unclear responsibilities. It is therefore for all parties involved to understand the contractual relationships, and for the lines of communication to be clearly stated.

CONTRACTS BETWEEN CLIENT, MAIN CONTRACTOR AND SUBCONTRACTOR

Irrespective of which procedure is used for appointing a subcontractor, it is always the task of the main contractor to place the formal contract. There are separate standard forms of contract for nominated or non-nominated subcontractors.

The client's nomination of a subcontractor implies an instruction to the main contractor to make a formal contract with the subcontractor.

As the main contractor is not responsible for design work carried out by the subcontractor, a special agreement between the client and the nominated subcontractor may be needed.

PAYMENT OF SUBCONTRACTORS

The main contractor is usually responsible for paying all subcontractors, including those nominated by the client. The same principles as were outlined earlier for contractors in general could be applied to the payment of subcontractors.

appointing suppliers

Most suppliers to a construction project are appointed by the main contractor, since he has the contractual responsibility for obtaining all the necessary materials for building. If the client wants a certain make of product, he may nominate suppliers. The same principles as for the nomination of subcontractors will then apply.

On government contracts in countries where there are shortages of basic materials such as steel and cement, the client ministry may supply these direct to the contractor. If this is the case, the terms should be clearly defined in the contract documents so as to avoid subsequent disputes. The client ministry may also wish to order directly certain specialised items, which are subsequently to be installed by the contractor. (An example would be sluicing sinks for a health centre.) Since such items usually have to be imported, government procedures may be lengthy.

It is one of the functions of the project manager to ensure that such items are ordered early enough to be available when required.

CONTROL 7

objectives

Control is an integral part of the project management process. It aims at the regular monitoring of achievement by comparison against planned progress. When deviations from planned progress occur, plans may have to be changed. Time is all-important, and the control process should aim at the early discovery of any departure from the planned course, so that adjustments can be made in time to be effective (see figure 25).

Figure 25. Control means monitoring progress and taking appropriate action.

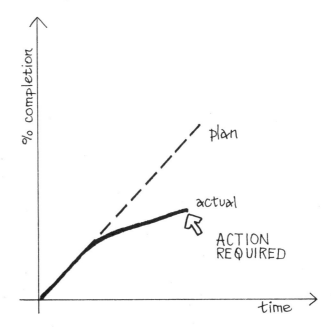

Figure 26. The control cycle.

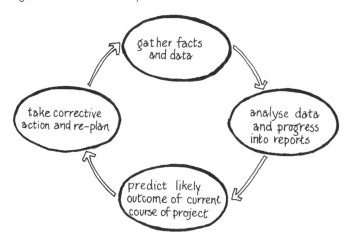

The control cycle is shown in figure 26. It is a continuous process throughout the life of the project, since it is very rare for any project to proceed exactly as planned.

Control information provides a basis for management decisions, and the following requirements should be satisfied by an effective control system:

- ☐ it should draw immediate attention to significant deviations from what is planned;
- ☐ true and meaningful comparisons must be possible;
- ☐ the information should indicate what corrective action is necessary, and by whom the action should be taken;
- ☐ it should be expressed in a simple form, so that it is readily understood by those who have to make use of it;
- ☐ key areas of control must be chosen with care, so that the results of control are worth the time and effort expended.

the role of the project manager

The project manager is responsible for overall control of the project. On a project large enough to justify a full-time management team, he may be assisted by a quantity surveyor, dealing with financial control, and a clerk of works, dealing with site progress and quality control. It may also be necessary to have specialists to help with the control of complex electrical installations or mechanical equipment.

The project manager often has to control by persuasion rather than by direct exercise of authority. This is one reason for the formation of management teams whose composition changes at each stage of the project. For instance, during the designing stage the architects working on the project may be in the Ministry of Works, while the project manager is in, say, the Ministry of Health. If the production of drawings has slipped behind schedule, the project manager cannot instruct the Ministry of Works to allocate extra resources to catch up. What he can and should do is demonstrate the effect of this particular delay on the overall project. If the architects have been involved in planning the project as part of the briefing and designing teams, the chances of a favourable response are greatly enhanced.

time, cost and quality control

There are three elements to be controlled in a construction project—progress against time; cost against tender or budget; quality against specification.

TIME CONTROL

As described in Chapter 5, the project manager should have prepared a time-schedule for the whole project during the briefing stage. Although lacking in detail, this will provide basic control information, such as the planned completion dates for each stage, as shown in figure 27. As the project progresses, information will become available from which a more detailed plan can be prepared. Thus, by the end of the briefing stage it should be possible to prepare a detailed work plan for the designing stage, showing not only when activities are to be completed but also the resources required.

Similarly it should be possible to prepare a more detailed schedule for the constructing stage well before

Figure 27. Outline control plan.

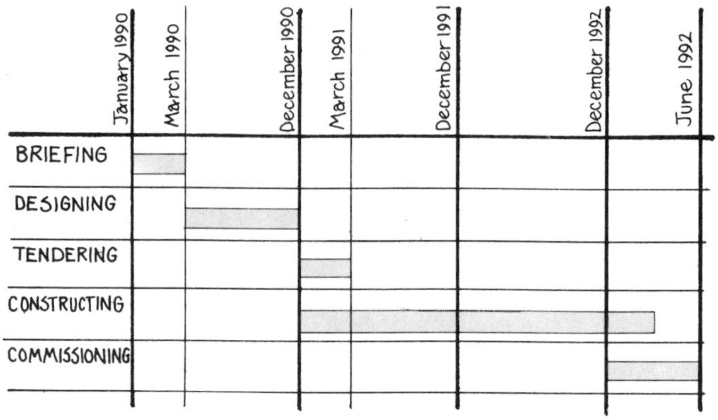

	January 1990	March 1990	December 1990	March 1991	December 1991	December 1992	June 1992
BRIEFING	▬						
DESIGNING		▬▬					
TENDERING				▬			
CONSTRUCTING				▬▬▬▬▬▬▬▬			
COMMISSIONING						▬	

completion of the designing stage, as shown in figure 28. Although the illustrations show bar charts, for large and complex projects it may be necessary to use network analysis. The required completion times for the constructing stage should be written in the contract document since they may have an important bearing on the tender price.

Figure 28. A construction programme.

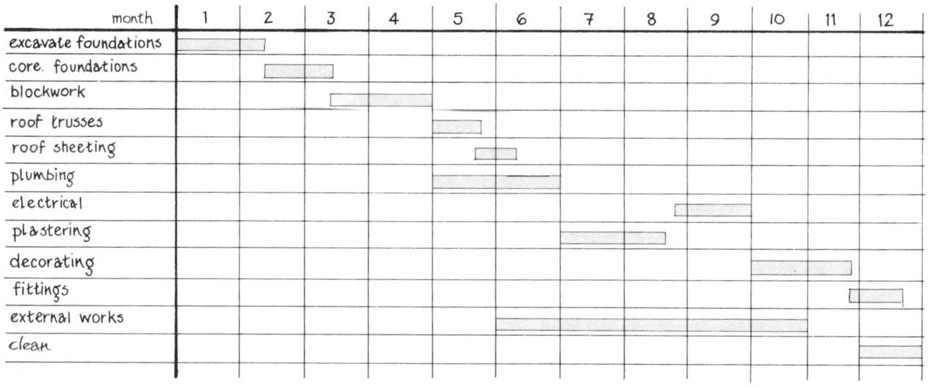

month	1	2	3	4	5	6	7	8	9	10	11	12
excavate foundations	▬											
core foundations		▬										
blockwork			▬									
roof trusses					▬							
roof sheeting						▬						
plumbing						▬▬						
electrical								▬				
plastering							▬▬					
decorating										▬▬		
fittings											▬	
external works							▬▬▬▬					
clean												▬

During the constructing stage it is the contractor's responsibility to prepare a detailed work plan which meets the requirements of the contract. The project manager should assess the realism of the contractor's schedules, especially as regards availability of resources. It is the responsibility of the main contractor to co-ordinate the work of his subcontractors; but there may be suppliers, or indeed other main contractors, whose activities the project manager must co-ordinate.

It is sometimes not clearly understood that the project manager does not usually have direct authority over the resources of men, materials and equipment needed to build the project. For example, if a brick wall is being built too slowly because the contractor has not employed enough bricklayers, the project manager cannot give instructions for the employment of more bricklayers. What he *can* and should do is point out the consequences of this slow building on the project as a whole. The principles and techniques of site management for those who do have direct control of resources would justify a separate volume, and are outside the scope of this Guide. However, a brief description of what is involved is given below under the heading "Site control".

The tools for progress control are the bar charts or critical path networks described in Chapter 5. Whichever technique is used, the project manager should take the following steps:

- ☐ establish "targets" or "milestones"—times by which identifiable complete sections of work must be completed. One such target would be the completion of sketch plans during the designing stage, or the completion of all works to render a building watertight so that equipment may be safely installed;
- ☐ as each target event occurs, compare actual against targeted performance. For example, were the sketch plans completed on the date planned, or two weeks before, or one month after?
- ☐ assess the effect of performance to date on future progress;
- ☐ if necessary, re-plan so as to achieve original targets or to come as near as possible to achieving them;
- ☐ request appropriate action from those directly responsible for the various activities.

Planning and control techniques achieve nothing unless they are translated into action, and it is the responsibility of the project manager to see that this happens.

COST CONTROL

The aim of cost control of a construction project should be the active control of the final cost to the client, not merely the passive registration of payments. As shown in figure 8, the possibility of influencing the final cost is greatest during the briefing stage and falls away rapidly when the constructing stage has been reached. The tool used for control is the project budget, and the principles shown in figure 29 apply to its use.

Figure 29. The project budget.

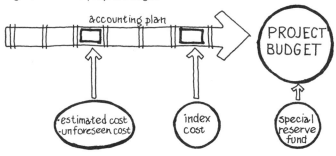

The project budget

A carefully prepared budget is vital for the effective control of project costs. It is also essential that it should be consistent with the aims of the project, and the required functions and quality standards. A decision on the project investment costs should be taken only after the preparation of a reliable cost estimate.

The project budget should be prepared according to an accounting plan. The choice of accounts and the level of detail will depend upon the client's information needs and the control requirements.

When the project is being executed for a private client, the budget should cover the complete project costs, including costs for the briefing and designing stages and general costs, such as service connection charges. The accounting plan serves as a check. The budget for each account is based on the estimated cost, with an addition for contingencies. The sum of individual budget accounts gives a total which is exclusive of price escalations and contingencies.

Price escalations can be dealt with in an account for "index costs" with a separate budget based on the net total, and a forecast of the likely percentage increase for inflation. The sums in the index budget are transferred step by step to the various accounts as resources are procured. This system allows budgets in individual accounts to be based on constant prices and so assists in realistic cost control.

To deal with general contingencies, a reserve fund may be budgeted for at the discretion of the client. This fund would cover such things as changes in the conditions upon which the budget was based, and would be used only with the client's specific approval.

Ideally, a government project should also observe the principles described above. However, government accounting procedures are designed for "steady state" activities and are generally ill suited to the control of projects. For instance, the services of architects and quantity surveyors may be provided by the Ministry of Works during the briefing and designing stages without any charge to the project. The same may apply to supervisory services supplied during construction. In such circumstances the project manager can do no more than set up a budget for those items over which he can exercise some control, which probably means the building works only. The way in which cost estimates are built up is now described. Since this Guide is concerned only with the project stages after a decision to build has been taken, no attempt is made to discuss how the initial "global" estimate is prepared.

Cost estimates

Several estimates of "cost" are prepared during the stages of a project. Care must be taken to distinguish between an estimate of "cost" and a "price": for example, a contractor may prepare an estimate of the cost to him of a project, but the tender price he submits will be his assessment of the highest figure that will secure him the contract. His *price* will then be the client's *cost*.

The reliability of any estimate depends on the information which is actually available at the stage when the estimate is prepared. Three categories of information are necessary—

- information about the project and its components. In the briefing stage information is normally restricted to rough estimates. Only when the sketch plans have been prepared is it possible to estimate costs with any degree of accuracy. A full knowledge of the quantities involved is available only when the working drawings have been finalised. Prior to this, however, it is possible to compile approximate "bills of quantities";
- information about resources. Only when the contractor starts planning site operations is information available about the manpower and other resources;
- information about prices. The contractor normally has a good knowledge of the actual prevailing prices for different materials and other resources. It is often more difficult for the designers to obtain reliable price information for the estimates prepared during the designing stage. Prices in "bills of quantities" from earlier projects can provide some information. During the briefing stage, the price information is normally limited to such data as the cost per unit of floor area, or unit length of road, as established from previously completed projects.

The choice of estimating method depends on the quality of the available information. An estimate of total construction costs proceeds through four main stages—

- preliminary estimates used in the briefing stage and based on cost records of broadly similar projects;
- detailed estimates, prepared by the project management team prior to tender, based on accurate quantities measured from working drawings and prices from previous project documents;
- contract sums, which may be a very good guide to the client's cost in the case of a fixed price contract, but less so in other circumstances;
- operational estimates, usually prepared by the contractor, based on the planning of site operations.

A cost estimate can never be more accurate than the information it is based on. Only an "operational estimate" or a fixed price contract sum can provide a really accurate cost estimate. The accuracy of earlier estimates is inevitably lower, with those at the briefing stage the least reliable. However, if one starts to produce estimates of

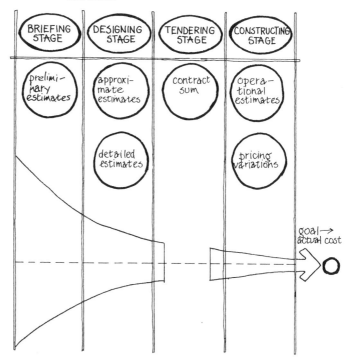

Figure 30. As estimating accuracy increases, deviation from "goal" decreases.

final costs at this very early stage, their accuracy can be progressively improved (see figure 30).

The forecast of the likely level of contractors' tenders will also require an assessment of the keenness of the competition, which will in turn vary with general economic conditions.

When submitting an estimate to the client at any stage, the project management team should always add a note on the probable degree of accuracy that has been achieved.

Cost control

Cost control should aim at ensuring that the final cost of the project does not exceed the budget. The greatest possibility for influencing the final project cost is in the briefing and designing stages. Regular cost checks should therefore be carried out on the developing design. A good aid in this work is a cost plan, based on an approximate cost estimate, indicating the quality, quantity and unit price for major cost elements such as earthworks, floors and roofs. When the design is developed in further detail, it is possible to check that the design of each element is kept within the framework set out in the cost plan.

An essential aid to cost control is a forecast of the final cost, which is regularly revised to reflect the current state of the project. If deviations between this forecast and the project budget are observed, corrective action must be taken.

A good aid to keep the cost forecasts up to date is a "cost diary" for each account, in which all events influencing the final cost are noted. These diaries should include such information as—

- ☐ cost checks prepared during the designing stage;
- ☐ contracts with consultants, contractors, suppliers and other organisations;
- ☐ variation orders, and variations foreseen;
- ☐ expected cost changes due to disturbances in the planned progress of the works;

Figure 31. The structure of cost monitoring and prediction.

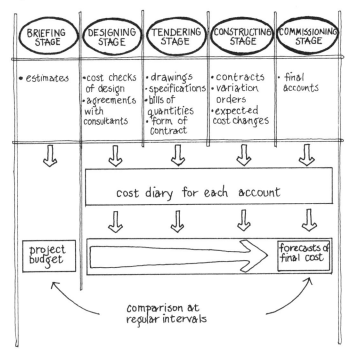

☐ differences between the actual and the indicated quantities and price fluctuations.

These elements are illustrated in figure 31.

Control curves

Once the contracts have been awarded, it is a fairly simple matter to estimate the "value" or expenditure on each activity shown on a detailed bar chart, such as that shown in figure 32. By analysing the nature of each activity, one can estimate the way in which this value will be distributed on, say, a monthly basis. These amounts may then be summed to give the total estimated monthly value, and the figures plotted in graphic form, as shown in figure 33. It is well known that the rate at which useful work is done is slower at the beginning and end of a project than in the middle, and thus these curves have a characteristic "S" shape.

Figure 32. Detailed bar chart.

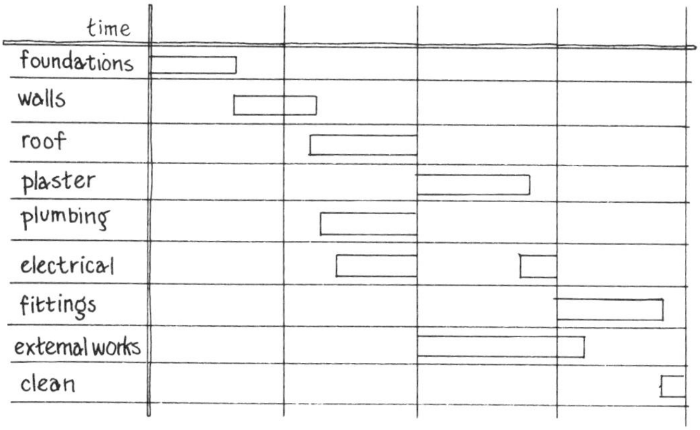

time				
foundations				
walls				
roof				
plaster				
plumbing				
electrical				
fittings				
external works				
clean				

If the work is valued as it is actually done, a second curve (shown as a broken line) may be drawn to indicate the cumulative value of the actual progress. In this way financial value may be used as a single, general measure of the progress of a project. This is shown in figure 33, where the "underspend" and overall delay are shown to be related.

Figure 33. "S" curve showing planned and actual values.

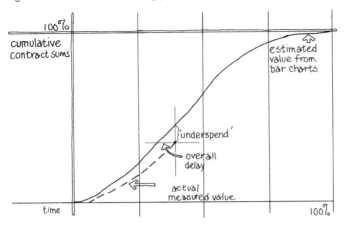

The continuous curves shown can of course be drawn only if it is possible to estimate the value of work done in detail. This implies a contract based on a bill of quantities. Where the contract payments are based on some other method, the project manager's ability to obtain a single and reasonably objective measure of overall progress is diminished.

Figure 34. "S" curves, applications and cash payments.

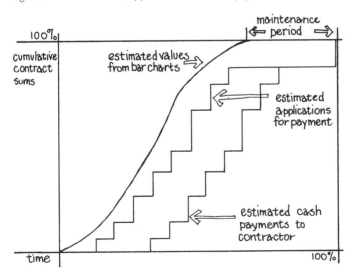

Cash flow

It is important to note that the S curves described above relate only to the value of work done, not to cash payments. The common system is that contractors are paid periodically, usually monthly, in a way related to the value of the work completed. The contractor submits an application for payment on a particular date each month. This is then checked and the project manager approves an amount which the client has to pay within a specified period. Thus the client's cash flow may be derived from the value S curve, as shown in figure 34.

QUALITY CONTROL

Quality control in a construction project should aim at satisfying the client's stated needs and requirements. Quality control must be exercised during all stages of a project, as shown in figure 35.

During the briefing and designing stages, a step-by-step decision-making process, incorporating "control stations" at which developing design is formally reviewed, makes the necessary overall control easier.

Quality control during the constructing stage is usually exercised on site by a clerk of works. He is responsible for seeing that the daily activities of the contractor result in an end product which satisfies the contract specifications. For example, he will ensure that the materials used for making concrete are up to standard, that they are mixed

Figure 35. Quality control throughout the project.

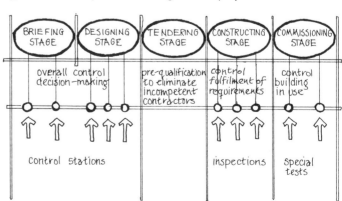

in the correct proportions, and that tests are made on samples of the mixed concrete. It may also be necessary for designers and specialists to carry out site inspections and checks. It is often useful to establish a control plan, as shown in figure 36, showing when, where and by whom checks are needed.

Figure 36. Use of control plan.

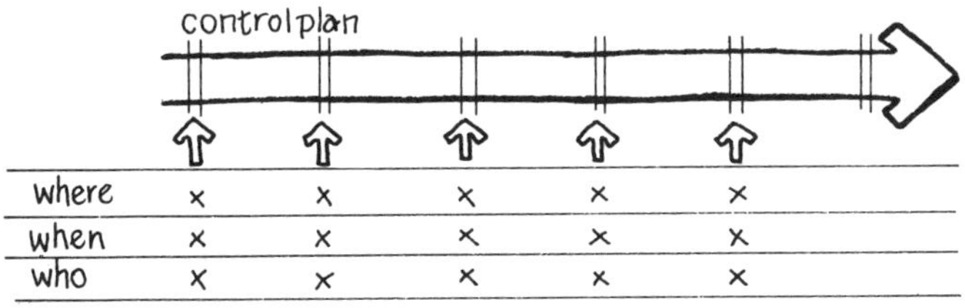

where	x	x	x	x	x
when	x	x	x	x	x
who	x	x	x	x	x

SITE CONTROL

The principal participants at the constructing stage are the project manager and the contractor. On a large contract the contractor may appoint a supervisor whom he may also call a project manager, so there is often confusion regarding the roles of these two main participants. As used in this Guide, "project manager" refers to the client's representative with overall responsibility for managing all stages of the project. "Contractor" refers to the person responsible for the construction works only. However, it is important that the project manager should understand the way in which a contractor manages a building project.

The project manager

The project manager can control site operations only within the conditions of contract. In the "standard approach", this means that he should record progress and notify the contractor if he is behind schedule. He can also, through the clerk of works, control quality against the specification. His control over the client's cost is limited to ensuring that variations from the contract drawings and specifications are minimised and that valuations are correct. Where a variation is unavoidable, it should be formally ordered on a standard form. The principles for pricing the variation order should always be agreed with the contractor beforehand. A fixed price is preferable.

Despite the apparent restrictions on his authority, the project manager can exert a considerable influence on site operations. Even on small projects there are many activities carried out by agencies and persons other than the main contractor. These activities should be included in the project manager's overall work plan, and he should co-ordinate them. He may also need to give technical advice to the contractor regarding the way in which the work should be carried out. This may be especially important on a small project where the contractor is inexperienced.

An important but often neglected aspect of site operations is that of safety. It is the prime responsibility of the contractor and other organisations working on site to ensure the safety of their employees. However, the project manager should see that the various regulations are being observed, and he should encourage the contractor to take common-sense safety measures such as keeping the site tidy. The next chapter describes safety in more detail.

The main site responsibilities of the project manager are as follows:

- ensuring the overall co-ordination of contractors, suppliers and other organisations working on site;
- obtaining the necessary statutory permissions. These may include site entry, building permits, work permits, notifications to safety inspectors, and so on;
- arranging for services such as water, electricity and gas;
- ensuring compliance with safety and fire regulations;
- monitoring progress and quality;
- checking periodic valuations and facilitating payments to the contractor.

These controls cannot be effectively exercised in an office remote from the site. On a large contract the project manager and his team must therefore be located on site. On smaller projects where the project manager may have a number of sites to look after, his site visits should be frequent and long enough to enable him to grasp fully what is going on.

The contractor

The word "contractor" is used to denote the organisation which actually erects the building and associated works. The principles of site control described here apply equally to parastatal contracting organisations, private contractors and direct labour forces, but the word "contractor" is applied to them all, for convenience.

The contractor directly controls the construction works. He assembles and organises the necessary resources of labour, materials and plant and equipment. Implicit in the supply of these is the need for a fourth resource: money.

The principles to be used by the contractor in managing these resources are exactly the same as those used by the project manager and shown in figure 26—

- ☐ gather facts;
- ☐ analyse the facts;
- ☐ predict the likely outcome;
- ☐ take corrective action.

Labour

The tender prepared by the contractor should have been based on his estimate of the work-hours of labour required to execute the project. Usually this estimate is built up from estimates of the required numbers of carpenters, masons, labourers, and so on. If he is to make a profit, the contractor must ensure that the productivity of his tradesmen and labourers is high enough to enable the work to be completed within the estimated work-hours. This can be done by measuring the work-hours for each activity or group of activities and comparing them with the estimated amount. If they exceed the allowance, the contractor must investigate the causes and take corrective action. If they are fewer, he must still check that the cost of these hours is not more than his estimate.

For example, the building of a roof may have been estimated to take 80 work-hours of carpenters' time, the rate for a carpenter being US$1 per hour. The total estimate is therefore US$80. If the work is completed in 74 hours, the contractor should be content—*unless,* in the period between the estimate being prepared and the roof being built, wages for carpenters have increased, say to US$1.40 per hour. The total cost will then be 1.4 × 74 = US$103.60. Whether or not the contractor can claim extra payment from the client depends on the conditions of the contract he has signed.

Materials

In the same way as the contractor estimates the cost of labour needed for the building, so he estimates the cost of materials. An important difference is that, apart from authorised variations, the quantity of materials is fixed. Whereas in the example given above the amount of labour needed to build the roof varies according to productivity, the quantity of timber, nails, roofing sheets and other materials is fixed by the drawings and specifications. It is still essential for the contractor to control closely his material usage, as building sites are very vulnerable to losses through wastage, theft or (in the case of such items as cement) deterioration.

Plant and equipment

These include items temporarily needed during construction, such as lorries, concrete mixers, hoists, ladders, buckets and shovels. The provision of these is the responsibility of the contractor and is included in his tender price. His responsibility does *not* include permanent building installations such as air-conditioning units, lifts, heating and ventilation units or generators.

Since all but the very large contractors cannot afford to own major and expensive items which may only be used for short periods on any one site, most contractors hire these as they are required. The same principles of measuring cost against estimate apply. Like labour, maximum plant productivity must be secured, and the hired items should be returned to the plant owner as soon as they are no longer required.

In the case of many government projects carried out by direct labour, plant is supplied from a plant pool which charges only the cost of fuel, maintenance and operators' wages. In these circumstances, there is little incentive to optimise productivity. Nevertheless, a good manager will release such scarce and expensive resources as soon as possible so that they can be used elsewhere.

Money

From the contractor's point of view, money is the most important resource of all. Without it he cannot acquire the other three resources he needs. That many contractors underestimate their need for this resource is demonstrated by the fact that in most countries they head the league table of bankruptcies.

What follows is *not* about profit, but about the working capital necessary to run a contract.

Even before building operations start, money is needed for—

- ☐ hire of supervisory and administrative staff;
- ☐ statutory fees, such as building permits;
- ☐ service connections;
- ☐ materials;
- ☐ hire of transport.

Many suppliers regard contractors as poor risks and are willing to accept orders only against a cash deposit. On some government contracts it is possible for a contractor to obtain a cash advance to help with these costs.

Many contractors underestimate their need for working capital. The fact that a certain value of work has been completed by a certain date does not necessarily mean that payment will be made soon after. On government projects particularly, bureaucratic procedures may delay payment for periods of three months or even longer. The contractor should be aware of this when preparing his tender. What this means is that he must have enough capital to continue paying wages, purchasing materials and hiring plant during the gap between a valuation of the work and actual payment. He must also bear in mind that it is usual for the client to retain 10 per cent or some similar figure of the valuation, to be placed in a special fund as an insurance that maintenance will be properly done. The effect of these delays and deductions is shown in figure 37. The working capital requirement is shown shaded.

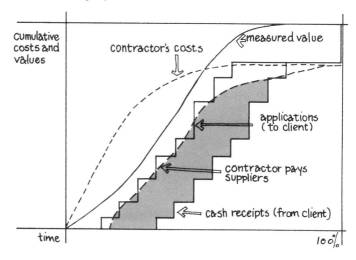

Figure 37. A contractor's control curve showing the need for working capital.

This need for working capital places a particularly heavy burden on small contractors, and so the project manager should do all he can to facilitate interim payments. A number of governments are taking active measures to deal with the financial and other problems of small contractors.

It is the usual practice for the contractor to be responsible for maintenance of his work for a fixed period (often six months) after it has been completed. To insure himself against failure to do so, the client retains 10 per cent of each interim payment up to a fixed amount. During the maintenance period the contractor must remedy all defects, which means that he may have expenditure but no income. This should also have been foreseen in his tender price.

The final certificate will normally be prepared by the project manager to certify that the work has been satisfactorily completed in accordance with the contract. Sometimes on government projects a technical audit unit prepares the certificate.

With the certificate there should be a final account which shows variations, including changes in quantities and price fluctuations, for which allowance has been made in arriving at the final contract sum. Only then can the final settlement be made with the contractor. This process sometimes takes several months, but it can be shortened if the project management team has kept proper records of all changes which occurred during construction.

HEALTH AND SAFETY 8

objectives

High standards of safety should be an objective pursued in the same way and with as much vigour as other management objectives.

The aim of most development projects is to improve the general well-being of the inhabitants of the country concerned. It is a reasonable humanitarian aim to ensure that the well-being of the people engaged in the project itself is preserved and perhaps enhanced.

In a construction project careful thought must be given to the health and safety aspects of the completed works, and the methods of site construction.

participants

Health and safety should be important considerations in all stages of a project, and so the project manager and the whole of the project management team are concerned. Indeed, they should encourage a general awareness and commitment from all parties involved in the project.

principal factors

The principal factors involved in the effective management of occupational health and safety are shown in figure 38.

Figure 38. Principal factors in effective health and safety management.

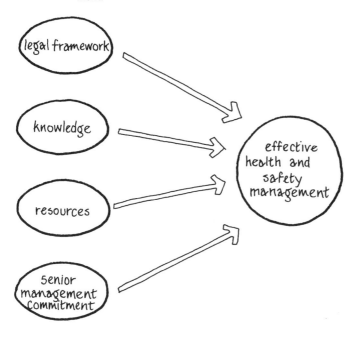

LEGAL FRAMEWORK

Most countries have a legal framework to ensure some degree of occupational health and safety. This legal framework usually lays down basic requirements of good employment practice, such as a minimum age for the employment of children. There may, in addition, be specific technical requirements relating to construction: for example, restrictions on the use of blue asbestos in buildings because of its detrimental effects on human health; or a simple technical construction requirement that the top of a ladder must be secured.

In some industrialised countries this legal framework has become extensive and detailed. In many developing countries it is rudimentary. Regardless of its level of sophistication and comprehensiveness, a legal framework can provide only a technical basis from which a coherent safety policy may be developed. Accidents and ill-health are not, as is so often believed, the result of straightforward technical failures; they result from a combination of social, organisational and technical problems.

THE NEED FOR KNOWLEDGE

Most people will take steps to reduce risk if they have sufficient knowledge of its existence. They need to know not just that the risk exists, but where, when and with what ferocity it will emerge. The key element is knowledge. The distribution and effective use of knowledge is a major management contribution to safety. With better information, instruction and training, most health and safety problems could be avoided. The ILO publication *Accident prevention* (Geneva, 2nd ed., 1983) provides valuable guide-lines in this respect.

The difficulty faced by managers is in making people fully aware of the need for safety. The key feature is the direct personal relevance of the information provided. General warnings, such as statutory warning notices displayed in works canteens, seem to have little effect. The project management team must take a positive approach to providing relevant, concise and clear information to the people involved, and do their utmost to ensure that this information is assimilated and acted upon.

Communication is considered in more depth in the next chapter.

RESOURCES

It is obvious that accidents and illness mean additional costs, and perhaps disruption of a project. It is, however, difficult to quantify these effects in financial terms, and equally difficult to quantify the financial benefits that may arise from the effective management of health and safety. Any measures taken which require the use of resources additional to the minimum required for "production" may therefore be seen only as additional expense. This expense may be significantly reduced if health and safety are given careful thought at the outset; but any safety policy must accept that some resources must be expended in achieving purely humanitarian objectives.

SENIOR MANAGEMENT COMMITMENT

Only senior management has the influence, power and resources to take initiatives and set standards. Positive attitudes of senior managers will be reflected in a high degree of health and safety awareness throughout the project. The converse is also true, and the lack of demonstrable interest by senior management in the welfare of the people involved will have a strongly detrimental effect on general morale and team spirit.

activities

Occupational health and safety can be considered under two headings—

- ☐ the completed (permanent) works, i.e. the activities in the briefing, designing and commissioning stages;
- ☐ the methods of site construction, i.e. the activities in the constructing stage.

The activities in the tendering stage provide some link between these two headings.

BRIEFING, DESIGNING AND COMMISSIONING STAGES

The relationships between occupational health and safety and the final form of the permanent, completed construction works are shown in figure 39.

Figure 39. Occupational health and safety and the completed works.

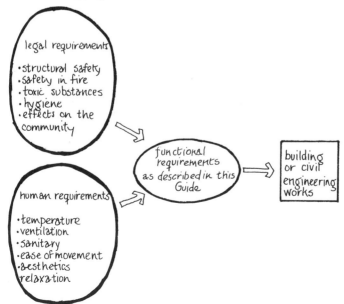

Legal requirements are made in most countries to ensure that the works are designed and built to minimum standards. Some of these are shown in figure 39, and examples may be given as follows:
- ☐ structural safety may be ensured by specifying the minimum dimensions and materials to be used in each structural element, or by requiring that the design conforms to a number of specified design standards and codes of practice;
- ☐ safety in the event of fire may be ensured by requirements relating to special fire exits and the use of specified non-combustible materials in important parts of the building, such as stairs;
- ☐ toxic substances may have to be stored in certain prescribed ways;
- ☐ hygiene requirements may specify a minimum number of lavatories for a building of a certain size intended to be used by a certain number of people, and may also require that these facilities be separated by some specific means from places where food is prepared or consumed;
- ☐ the requirements for the design and construction of large dams are usually very demanding, because their failure may have a catastrophic effect on the community down river.

Human requirements are also shown in figure 39 and need no further explanation.

CONSTRUCTING STAGE

Accidents on construction sites may occur in any of the following ways:
- ☐ through the collapse of walls, building parts, stacks of materials, masses of earth;
- ☐ through the collapse or overturning of ladders, scaffolds, stairs, beams;
- ☐ by falls of objects, tools, pieces of work;
- ☐ by falls of persons from ladders, stairs, roofs, scaffolds, buildings; through hatches and windows; into openings;
- ☐ during loading, unloading, lifting, carrying and transporting loads;

- ☐ on or in connection with vehicles of all kinds;
- ☐ at power plant and power transmission machinery;
- ☐ in the operation of railways;
- ☐ on lifting appliances;
- ☐ on welding and cutting equipment;
- ☐ on compressed-air equipment;
- ☐ by combustible, hot or corrosive materials;
- ☐ by dangerous gases;
- ☐ during blasting with explosives;
- ☐ when using hand tools.

Occupational diseases of construction workers include—

- ☐ silicosis (stonemasons, sandblasters);
- ☐ lead poisoning (painters);
- ☐ diseases of the joints and bones (workers operating percussive tools);
- ☐ skin diseases caused by such materials as cement.

SAFETY AND LABOUR-BASED TECHNOLOGY

Large construction projects in developing countries may lead to a disruption of a traditional way of life for temporary employees. Problems of transport, living, working, training and safety may be accentuated. There will be problems in giving new workers an understanding of the more elementary principles of safety in this new environment, particularly in the use of machinery. Many of these problems may be avoided if the works have been designed to permit the use of traditional practices and tools with which the workers are already familiar.

During the briefing, designing and commissioning stages the project manager and his team can control the occupational health and safety aspects of a project closely and directly. They are in direct control of the parties involved, and their main task will be to ensure that diligent attention and commitment are given to this aspect. The main problems are likely to be cost, and complexity of requirements.

causes of accidents

There are as many possible causes of accidents as there are occasions. Among these are technical defects in equipment and methods of work, defects in organisation and dangerous acts by workers. To these have to be added those causes that come from the nature of construction operations themselves: defects in planning and construction; constant changes in workplace and task; and the friction often found when workers from different trades are working in close proximity to each other. In the following list, the causes of accidents have been grouped according to their nature.

Planning, organisation
- ☐ defects in technical planning;
- ☐ fixing unsuitable time-limits;
- ☐ assignment of work to incompetent contractors;
- ☐ insufficient or defective supervision of the work;
- ☐ lack of co-operation between different trades.

Execution of the work
- ☐ constructional defects;
- ☐ use of unsuitable materials;
- ☐ defective processing of materials.

Equipment
- ☐ lack of equipment;
- ☐ unsuitable equipment;
- ☐ defects in equipment;
- ☐ lack of safety devices or measures.

Management and conduct of the work
- ☐ inadequate preparation of work;
- ☐ inadequate examination of equipment;
- ☐ imprecise or inadequate instructions from supervisor;
- ☐ unskilled or untrained operatives;
- ☐ inadequate supervision.

Workers' behaviour

- ☐ irresponsible acts;
- ☐ unauthorised acts;
- ☐ carelessness.

project management team functions

During the construction stage the directness of this control will depend upon the contractual arrangements. If the construction is to be done by direct labour, the project manager and the construction team will have direct control, and the only problems will be financial and technical.

If the construction work has been let to a contractor, the responsibility for site safety passes to this contractor. Some control may be exercised, if safety requirements are included in the contract: it is usual, for example, to state that the contractor shall comply with all relevant local legislation. In addition, the contractor is usually required to take out insurances to protect the public, as well as the people employed on the site. Insurances, however, merely compensate for injury—they do not prevent it. Where a workers' union exists, the contractor should consult with the union in implementing safety measures.

It is important that the contractor be made clearly responsible for practical health and safety provisions on site. If the project management team intervenes and issues direct instructions for the specific provision of safety measures, the client may be held liable for their cost, and also may be held responsible if they are inadequate and cause injury. The project management team can influence safety on site beyond the contractual minima only by means of persuasion.

COMMUNICATION AND REPORTING 9

Management relies on clear communications, and the ability to pass thoughts, ideas, information and instructions quickly and effectively between people with different technical skills and interests.

managing people

Management is about translating objectives into reality. This can only be done through people. The main character in this Guide is the project manager, and it is therefore important that he should understand how to deal with people. He operates in different ways as the project proceeds through its various stages. His direct authority does not extend beyond the full-time members of his team. His authority over those who join the team from time to time is moral rather than absolute. The success of the project therefore depends a great deal on the personality of the project manager, and on his ability to persuade others to perform their functions as and when required.

Good communication between managers and staff of the various organisations involved in the project helps to create enthusiasm where it is most needed. In this connection it is the responsibility of the contractor to establish good communications with his workers; and where a union exists, there should be full consultation at the start of a project. Bad or inadequate communication produces indifference and even antipathy. Much valuable staff effort is then directed towards overcoming this break in communication, instead of being spent more productively on solving more important problems. To achieve good communication it is necessary to create an appropriate and workable administrative organisation.

In any administrative organisation it is necessary to define the limits of each person's (or group's) responsibilities. Each tier or position in an administrative hierarchy (see figure 40) has a frame of actions to perform, and certain limits of responsibility. One common misapprehension is that this hierarchy also defines the system of communications, and this, in turn, leads to a mistaken belief that direct communication between people not directly linked in the hierarchy is somehow bad.

Figure 40. Formal communications work through the organisational hierarchy.

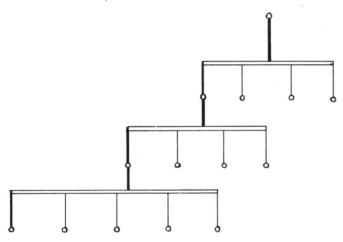

For simplicity we may say that effective communications occur in two ways: informal and formal. Informal communications are valuable for establishing good personal relationships, for the speedy and effective resolution of problems and for deciding upon courses of action. Formal communications are required to ratify a decision made informally, to record briefly the main reasons for this decision, and to communicate the relevant information to people who were not involved in making the decision. Formal communications are also necessary for the orderly conduct of a project: applications for funds, certification and payments; and periodic reports and accounts to aid donors or government agencies.

INFORMAL COMMUNICATIONS

Good relations between people are the core of effective management. The ability to handle human resources properly will be reflected directly in staff morale, team effectiveness, productivity and project efficiency. Much more can be achieved by a few people with a sense of common purpose than by a large organisation lacking drive or effective management.

The management approach must be adapted to the size and complexity of both the project and the administrative organisation. In a small organisation informal communications are easier, since people find their own ways to keep informed. In a large organisation formal communications are more necessary so that everyone who needs to keep in touch is able to do so. This is shown in figure 41.

Figure 41. Formal and informal communications.

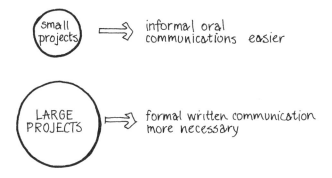

There can be no doubt that the most efficient way to communicate information from one person to another is orally, face to face. This form of communication is more efficient because it does not just rely on words: gestures, eye contact and other forms of "non-verbal communication" contribute to establishing the true meaning and intention of the people taking part (see figure 42). In addition, the communication process relies on interaction between the parties, and in the case of oral communication this may be instantaneous, resulting in speedy action. Other forms of informal communication include telephone conversations and handwritten notes, both of

Figure 42. Face-to-face communication.

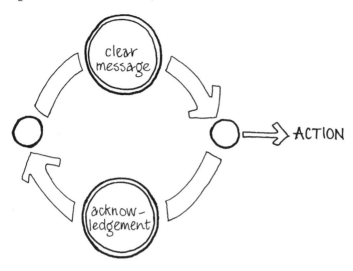

which are very effective. Typewritten notes appear more formal, so that many managers use them only when they wish to formalise communication.

FORMAL COMMUNICATIONS

Financial accounts are the most obvious example of formal communications. Their purpose is to show that the financial affairs of the organisation have been soundly and honestly managed. To avoid ambiguity and to reduce the risk of dishonest manipulations, financial accounts are prepared and checked according to a set of well-established and publicly available procedures. Nevertheless, there is scope for simplification in most such systems.

Many of the formal reports and procedures used in a project exist for similar reasons and are prepared in a standard way. They are necessary and should be done diligently and efficiently.

The drawings, specification and other contract documents used to define the project are other necessary forms of formal communication. The purpose of these documents is to define the work clearly and simply, in such a way that—

☐ the people involved in producing the physical end product can understand what has to be done, and by when;

☐ the people supervising the work can easily understand whether the end product is satisfactory or not.

Unfortunately, most of the project documents used in developing countries are far from clear and simple. This is because they have been derived from models used in industrialised countries, with little or no adaptation to local conditions. Contract documents, which are legally enforceable, should not be difficult to understand; but in many countries there is room for a great deal of improvement in this important respect.

STRIKING THE RIGHT BALANCE

The ability to control communications is one of the skills required by a project manager. Although oral communication is effective, there is danger in placing too much reliance on the spoken word, especially where the repercussions of a spontaneous remark have not been thoroughly thought through, and no record exists. Conversely, flexibility and time may be lost if every action or decision is delayed until a written communication or confirmation is received.

meetings

Meetings and minutes are an important link between informal and formal communications. They enable a number of people to be informed, to exchange information, and to reach decisions simultaneously.

Meetings range from simple conversations and get-togethers to conferences, subcommittees, working parties and project meetings.

When he calls a meeting, a senior manager will (if he is wise) encourage a free expression of views. However, when all has been said, he alone must take the final decisions, and he will be held responsible for them. In contrast, the chairman of a committee is charged with the efficient running of the meetings. It is the members of the committee who reach the decisions and who are responsible for these. All the members of a committee bear equal responsibility for the decisions reached, and individuals can dissociate themselves from majority decisions only by resigning.

Conducting and taking part in meetings of all kinds is itself an important management skill. For meetings to be successful and effective, highly developed social skills must be brought into play, sound administrative arrangements made, and a proper attitude to participation cultivated. Three phases can be identified for every meeting or committee: planning, conduct and follow-through.

PLANNING THE MEETING

First, if it is to be effective, the precise purpose of the meeting must be clearly stated. This must be done in advance, so that those who are to take part understand what contributions are expected, and therefore come prepared to make them. Second, there should be a policy of "no business—no meeting". Third, when the purpose for which the meetings were convened has been outlived, the meetings should be abandoned.

The number of people at meetings should be as few as possible, with proper use made of subcommittees and working parties. Advisers, experts and observers should be involved as necessary, but they need not be recruited as members. All these people must be brought together, on the same day, at the same time, and in the same place, to talk about the same things in the same order. This means giving adequate notice and making suitable arrangements. In order that decisions may be based on due reflection, it is also helpful if material which is to be considered at the meeting is circulated beforehand, and not tabled during the course of the meeting.

CONDUCTING THE MEETING

The key man in any meeting is the chairman. He should be positive in his approach, he should know the business of the meeting beforehand, and he should hear and be heard by everybody. His role is to—

- ☐ put propositions and points for deliberation, explaining clearly the topic to be considered;

- adopt a systematic approach, allowing full discussion but avoiding irrelevances and meandering;
- listen for what is *not* said, stimulating members to brisk performances within an allotted time;
- ensure that any resolution put to the meeting is clearly worded so that it will stand the best chance of securing general understanding; and
- summarise what has been agreed, and the action to be taken, and by whom, before moving on to the next item.

At the end of the meeting, there must be no doubt about what has been decided, what subsequent action is required, and by whom. Much of the value of a meeting may be lost if steps are not taken to ensure that decisions are properly implemented.

FOLLOW-THROUGH

It is always necessary to confirm what was agreed at the meeting by circulating "minutes"—taken by an appointed secretary—as soon as possible afterwards. Ideally, everyone present should also take notes. Minutes should include a record of all decisions made during the meeting, together with supporting facts. A précis of the discussion should be avoided. Minutes should be issued as final, and subject only to correction of any error of fact and to any agreed amendment.

To ensure that the responsibility for the action which is to be taken is clear, an "action" column can form part of the minutes. In this the initials of the person concerned can be noted.

Communication, it should be remembered, is a two-way process, and its form must be modified to suit the circumstances and the size of audience.

summary

We may summarise project communications as follows:

- informal, oral, face-to-face communication is effective and quick. However, any decisions reached should

be confirmed in writing and other interested parties kept informed;
- ☐ the organisation chart (figure 40) does not depict a system of communication; but anyone communicating *across* the hierarchy must observe the limits of his authority and responsibility—and must also inform people further down the hierarchy;
- ☐ drawings and other documents that define the physical work must be clear, direct and immediately relevant to the conditions and environment in which they will be used;
- ☐ formal communications such as financial reports must be done as efficiently as possible. These systems should be designed to be as simple as possible, particularly at the operating level;
- ☐ meetings, even simple meetings between two or three people, serve as an important link between formal and informal communications.

PLANNING TECHNIQUES   10

This chapter describes three planning techniques: time-chainage charts, network analysis, and bar or Gantt charts. The techniques are described by using examples to demonstrate their application to construction projects in development programmes.

The planner's first task is to select the right technique for the job. Figure 43 illustrates a selection procedure. The job may be a whole project or a part of that project. For example, a complex project may be planned and controlled using network analysis; but a road, which may be represented by a single activity ("build road") in the network plan, may be planned in more detail using a time-chainage chart. Similarly, a very straightforward, simple project may require no more than a bar chart for overall control but some small parts may benefit from network analysis at the detailed level.

Figure 43. The selection of planning techniques.

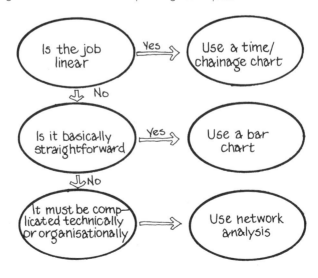

time-chainage charts

Time-chainage charts are simple graphs of the project plan, where distance (or "chainage") is plotted on the *x* axis and time on the *y* axis. The *x* axis thus represents the linear job being planned.

Figure 44. Construction of a simple earth road.

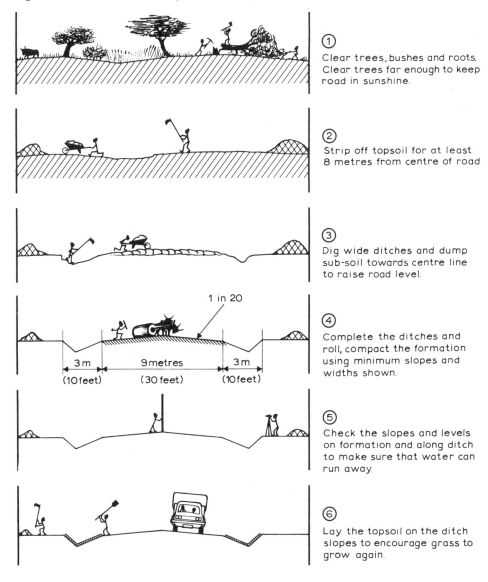

Source: M. Allal and G. A. Edmonds, in collaboration with A. S. Bhalla: *Manual on the planning of labour-intensive road construction* (Geneva, ILO, 1977).

A good example is that of the construction of a simple earth road. The basic construction is illustrated in figure 44, and a plan and section of the proposed road are shown in figure 45. It will be seen that a stream crosses the line of the proposed road, that this is to be conducted through a new culvert, and that the small rise to the west of the stream is to be lowered and the excavated material used to fill the ground to the east. The stream carries water only during the rainy season.

Figure 45. Plan and section of a simple earth road.

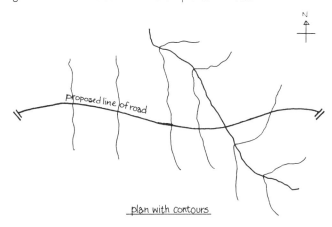

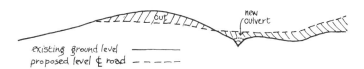

The contractor's time-chainage chart is illustrated in figure 46. The culvert will obviously be built at a fixed chainage and must be completed before the rains come, so that construction work is done before the stream floods. The main earthworks will not begin until the culvert has been completed, because the dry stream bed is very rough and has steep banks. When the culvert is complete, excavation will begin immediately to the west

Figure 46. Time-chainage chart for road shown in figure 45.

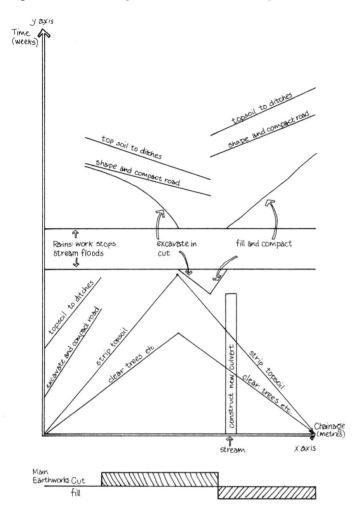

of the culvert and the excavated material will be placed and compacted immediately to the east of it. In this way, the animal-drawn carts moving the spoil will always be travelling on the relatively flat gradients of the new road, and so will work more efficiently. The excavation line is curved because the excavation is deeper in the middle of the cut, and progress *along* the road is therefore slower. The depth of fill is, however, fairly uniform.

Work to the west of the cut area will be unaffected by the progress on the culvert, and so can start immediately. It is planned to complete all work on this section before the rains, so that the grass sown in the ditches may benefit from the water.

Figure 44 shows that clearing trees and bushes, and stripping the topsoil, takes place across the road, and this also is therefore unaffected by the culvert. Two teams will be employed on both these operations, starting at the ends of the road and meeting in the middle.

network analysis

("ACTIVITY-ON-NODE" OR ("PRECEDENCE" METHOD)

The example chosen for network analysis is of a type of timber truss bridge that has been developed in Kenya. A typical bridge is shown in figures 47 and 48.[1]

The objective was to provide relatively cheap bridges to carry light commercial vehicles in rural areas. The

Figure 47. The Kenyan low cost modular timber bridge.

Source. J. D. Parry: *The Kenyan low cost modular timber bridge*, TRRL Laboratory Report 970 (Crowthorne, Berkshire, United Kingdom, Overseas Unit, Transport and Road Research Laboratory, 1981). © Crown Copyright 1981. Transport and Road Research Laboratory. Reproduced by permission of Her Britannic Majesty's Stationery Office.

Figure 48. Cross-section of typical bridge.

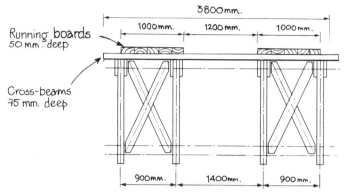

Source. Parry, op. cit. © Crown Copyright 1981. Transport and Road Research Laboratory. Reproduced by permission of Her Britannic Majesty's Stationery Office.

bridge comprises a number of identical timber frames that are assembled into trusses of the required span. Two or more parallel trusses are supported on conventional abutments, and the timber deck rests on top of the trusses. The erection procedure is shown in figure 49. This design has the additional advantage that the bridges can be

Figure 49. Erection of a bridge.

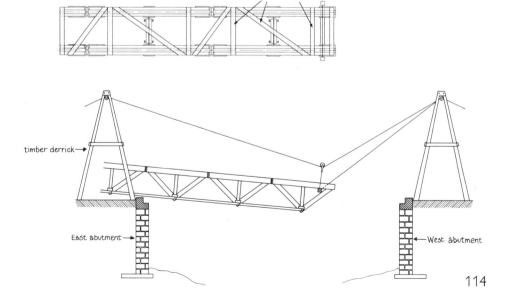

erected quickly, and can be dismantled and re-erected at another site if required.

The provision of the bridge is simple, and planning could be done quite well with a bar chart. It is, nevertheless, a useful project on which to demonstrate network analysis because the sequence of the activities is fairly clear.

As the "activity-on-arrow" method was explained earlier in the Guide, this more detailed explanation of network analysis will begin with the "activity-on-node" or "precedence" method.

Activities

The first task in planning is to identify the activities necessary to complete a project. The advantage of network analysis over other techniques is that this may be done as a preliminary process, unrelated to considerations of resources, durations and inter-relationships between activities.

In precedence networks the activities are depicted by boxes, usually rectangular, as shown in figure 50. The activity is briefly described within the box, and may be given a number for ease of identification.

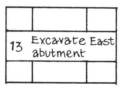

Note that if the network becomes too long for the page it may be stopped at some convenient point and re-started elsewhere. The links are provided by repeating the activities (in this case 9 and 12) and emphasising this with an extra line round each box.

Constraints and activity relationships

The logical constraints and the relationships between activities may now be considered. These are expressed as arrows as shown: for example, it is impossible to start activity 5 ("cut and form timber for frames") until activity

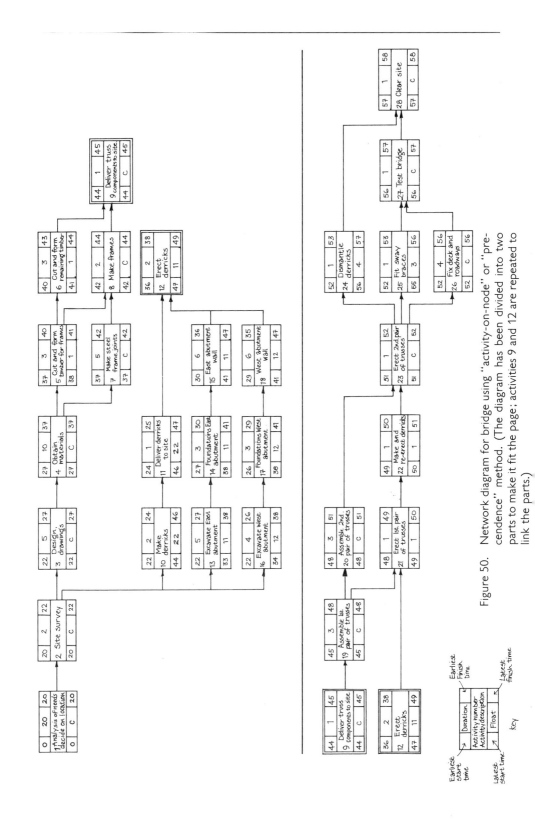

Figure 50. Network diagram for bridge using "activity-on-node" or "precendence" method. (The diagram has been divided into two parts to make it fit the page; activities 9 and 12 are repeated to link the parts.)

4 ("obtain materials") is complete. Another example is that it is impossible to construct an abutment wall foundation before the necessary excavation is complete (i.e. activity 14 depends upon the completion of activity 13).

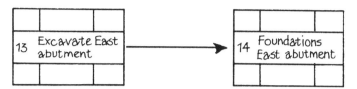

By following reasoning of this kind it is a relatively simple task to prepare a diagram similar to figure 50.

It is possible to develop these logical constraints further. Taking the example of activities 13 and 14, it may be possible to begin the foundations before the excavation is 100 per cent complete—perhaps after two days' work. Thus activity 14 may start two days after activity 13 has started; this is known as a "start-to-start" relationship and is drawn as shown:

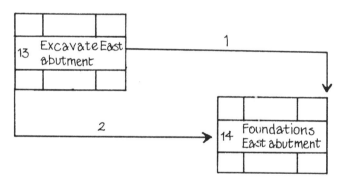

In order to complete the foundations it may be that the excavation must be finished in time to allow one complete day for foundation work. This would be a "finish-to-finish" relationship.

These devices allow the network diagram to reflect reality more closely, but they result in more complex analysis. In many cases it is doubtful whether this extra effort really gives better planning and control, and therefore their use is not generally advocated in this Guide.

Duration of activities

When the network diagram has been drawn, the expected duration of each activity may be added. At this stage the durations should be those expected from the resources commonly employed on the particular activity, operating at an average efficiency. The argument for this is that the critical path analysis which follows identifies those activities where a reduced activity duration will result in a corresponding reduction in project duration. It is on these critical activities that additional resources and intensive working should be concentrated.

The duration of each activity should be estimated by people who have experience in managing activities of that type. In most cases they will rely on records of the progress of similar activities in the past, and they will use their own judgement and experience to apply these data to the activity under consideration. For example, a senior architect would estimate the time it would take his staff to produce the drawings for a new building by examining his records of the number of drawings that were required to describe comparable buildings in the past, together with records of the time taken to produce drawings of various types. The estimated activity duration would be the result of the largely arithmetical process; but the architect would then adjust this estimate to reflect his own assessment of the design difficulties likely to arise, and the ability of his staff.

Critical path analysis

When the durations have been added, it is possible to calculate the "critical path": that is, the sequence of activities which determines the overall job duration. Any delay in any activity on the critical path will result in a similar delay to the whole job. This analysis will also identify by how much each activity may be delayed before it too results in an overall delay in the completion of the job.

There are three parts to the analysis—

- ☐ the forward pass, which determines the earliest times that every activity may start and finish;
- ☐ the backward pass, which determines the latest times that every activity may start and finish;
- ☐ the calculation of "float", i.e. the time available to each activity for delay before other activities become affected.

The forward pass

This begins by establishing the time for the job to start. This may be a calendar date, or a week or day number if the project calendar is expressed in that way. In figure 50 the job (in this case a complete project) begins at the beginning of day zero with a general analysis of the needs for the bridge and a decision upon its approximate location. This first activity has a duration of 20 days; therefore its earliest finish is 0 + 20 = 20, i.e. the end of day 20. This activity is followed by activity 2, which cannot begin until the first activity is complete. i.e. day 20. With a duration of two days, activity 2 will have an earliest finish time of 22 days. (These calculations obviously refer to working days, and so rest days and holidays must be included before calendar dates can be determined.)

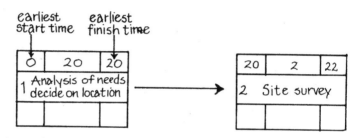

The forward pass proceeds through the network in this way, but there is a complication if the activity under consideration is *preceded* by more than one activity; for example, 8 is preceded by both 5 and 7. Activity 5 has an earliest finish time of 40 days, whereas activity 7 finishes after 42 days. Since the logical relationships expressed by the arrows show that *both* 5 and 7 must be complete before 8 may start, *the rule is to take the latest time on the forward pass when there is more than one preceding activity*: in this case, 42 days.

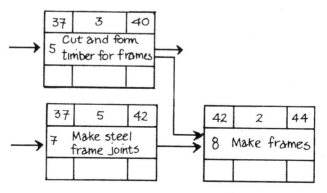

At the end of the forward pass the estimated time required to complete this project, according to the plan expressed in the network, is established: in this case, 58 days.

The backward pass

The *latest* times for activities to start and finish may now be established by means of the backward pass. This is an essential step in establishing how critical the timing of each activity is to the overall duration of the project. If an activity's earliest start date (i.e. the earliest time when it *can* start) is the same as its latest start date (i.e. the latest time by which it *must* start), this activity is critical.

The backward pass is simply a reversal of the forward pass, and calculations begin at the last activity. Complications again arise when several activities are related, but in this case a difficulty arises when an activity has more than one *succeeding* (that is, dependent) activity, as in the case of 5 followed by 6 and 8. Working backwards through the network, 6 has a latest start time of 41 days and 8 has a latest start time of 42 days. The logic of the diagram says that *both* 6 and 8 are dependent upon the completion of 5, so 5 *must* be complete by the *earliest* of the two times, i.e. 41 days, not 42. *The rule is to take the earliest time on the backward pass when there is more than one succeeding activity* (this is therefore a direct reversal of the forward pass rule).

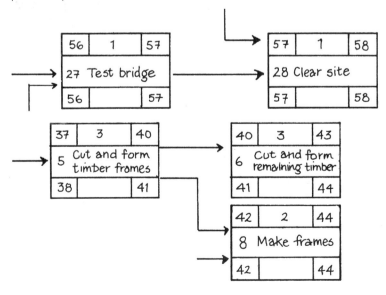

This analysis proceeds backwards through the network, and should result in the earliest and latest start times for the first activity being the same.

Float

"Float" is the time available to an activity for it to "float" between the earliest time that it *can* start and the latest time by which it *must* start if the completion date of the job or project is not to be delayed.

Total float is the total float time available to an activity: that is, its latest finish less its earliest start time, less, of course, its duration. This means that it is its latest start time (or finish) less its earliest start time (or finish).

Float is an important means of controlling projects. When resources are inadequate for all available activities, those with float can be temporarily suspended without delaying the overall project time.

For activity 13 the total float is 33 − 22 = 11 days. Thus, if activity 13 starts on any of the 11 days between these two dates it will have no effect on *the overall project completion date.* If this total float had been zero, the activity would have been critical and a "C" would have been entered in the box. The "critical path" is this chain of critical activities, and it is this chain which will be the focus of the project manager's attention. (There may, of course, be projects with more than one critical path.)

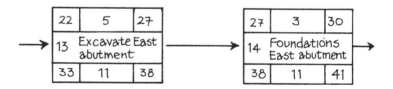

The calculation of total float includes the time by which any subsequent activities may be delayed (without, of course, extending the overall job duration). For example, the calculation of total float for activity 13 used the latest start time (33 days) from which was deducted its earliest start (22 days). However, if activity 13 does not start until 33 days have expired, it will not be complete until the end of day 38 (i.e. 33 + 5). Activity 14 will then be *forced* to start at its *latest* start time (day 38) and in this case it has lost its float to activity 13. Similarly, activities

16, 17 and 18 have 12 days total float, but it can be seen that they are all part of a "chain" of activities. A "chain" is a simple string of activities with single "finish-to-start" dependencies, and they have this characteristic that all the activities in the chain "share" the total float.

This may be emphasised by comparing activities 13, 14 and 15, which all have a total float of 11, with activities 16, 17 and 18, with a total float of 12. It is obvious that both these chains contain similar activities, except that less excavation is required for the west abutments and activity 16 therefore takes one day less than activity 13. This "extra" day nevertheless appears in all total float of all the activities in the chain.

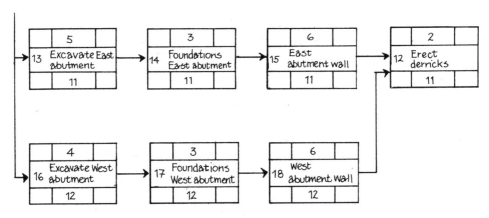

It is clear, therefore, that total float is an important part of planning analysis and is easy to calculate. Its interpretation does, unfortunately, require some care.

Free float provides a partial solution to the problem. It is the time by which an activity may be delayed beyond its earliest start time without delaying *any* subsequent activity beyond its *earliest* start time. It is clear from figure 50 that activity 13 has no free float, because if it is delayed beyond its earliest start (22) it would not be complete by its earliest finish (22 + 5 = 27) and consequently would delay activity 14 beyond its earliest start (27). Similar arguments apply to 14/15, 16/17 and elsewhere. Activity 18, however, is at the end of its chain and could be delayed by one day without delaying activity 12 beyond its earliest start (35 compared to 36). Thus activity 18 has one day free float. Because the use of free float has no effect on succeeding activities, it is a most useful facet of critical path analysis.

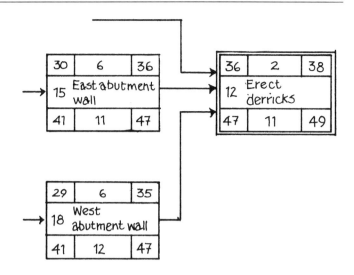

"ACTIVITY-ON-ARROW" METHOD

"Activity-on-arrow" is an alternative method of drawing the network, and it has been described in general terms earlier in this Guide. Figure 51 shows an "arrow" form of the precedence diagram shown in figure 50. The advantages and disadvantages of the two methods are fairly evenly balanced, and the decision which to use is often based on the system with which a majority of the project staff is familiar.

The principles are fundamentally similar. The main difference is that the arrows in the activity-on-arrow method represent both activities and the activity relationships. The maximum number of activity relationships therefore becomes related to the number of activities. This is clearly impractical, and additional logical relationships therefore have to be introduced in the form of "dummy activities" (shown dotted). These are not required to represent actual project activities, but serve solely to represent activity relationships in a similar way to the arrows in a precedence diagram.

The activities are linked by circles, called "events". These signify that all the preceding activities have been completed and that all subsequent activities may start. The critical path analysis is based on exactly the same principles for both arrow and precedence diagrams, but

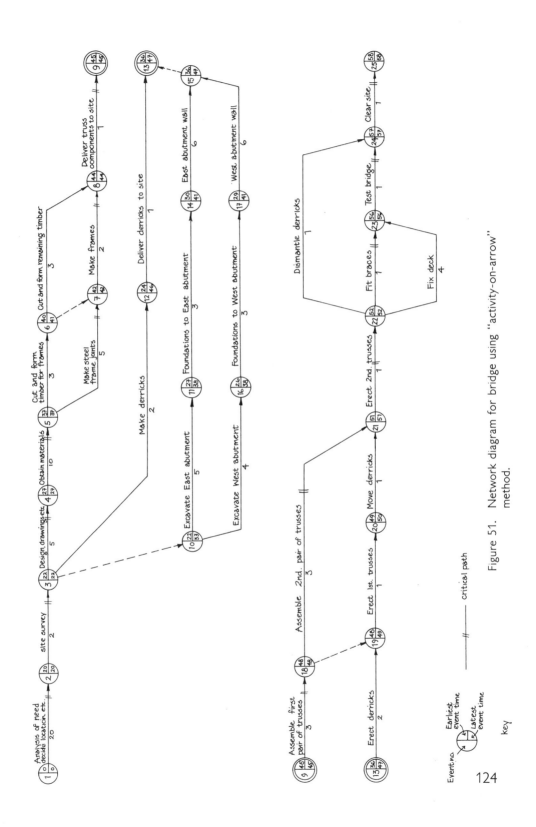

Figure 51. Network diagram for bridge using "activity-on-arrow" method.

the arrow diagram causes some difficulties of understanding that the precedence diagram avoids.

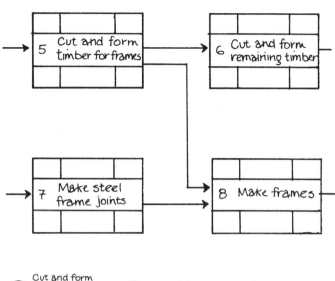

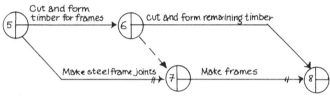

NETWORK ANALYSIS IN PRACTICE

Network analysis should not be seen solely as an analytical technique. Its great power is that it can form a disciplined pictorial focus for project management team discussions. The first draft of any network diagram should evolve during the discussions of the people concerned; it should not be produced by a specialist planner working in isolation. As the team works through the plan and draws the network, *all* members of the team gain an understanding of the attitudes of the other members of the team, and of the problems they are likely to encounter. A specialist planner may be present at the meeting to help with the mechanics of the network, but he should not be central to the decision-making process.

Another advantage of this approach to network analysis is that continuing control of the project becomes easier. When a problem arises, the team members can quickly comprehend its likely effects on the whole project and re-plan accordingly.

All this implies a simplified approach to network analysis, and it is for this reason that the explanation given in this Guide is much more simple than that given in many academic textbooks. For example, no suggestion has been made that float should be calculated by listing all the activities and their relevant timings, and then performing the calculations using mathematical formulae. Instead, a basic description of what "float" means has been given. There is not time in a management meeting to list all the activities, and in any case this is unnecessary because the attention of the members of the team will be focused on a few activities only. What is required is the ability to calculate quickly the total float available to the relevant activities, to understand what this means (e.g. are they in a chain?) and to decide whether or not it is wise to allow the activity to be delayed.

bar or Gantt charts

Bar charts were described earlier in this Guide. It is also possible to draw a bar chart from a network diagram. Figure 52 shows the network of figure 50 in bar-chart form. This has been prepared simply by plotting the earliest start time of each activity, followed by a bar to represent its duration. If the latest finish time is also plotted, the total float may be obtained graphically, as shown. Free float may also be obtained graphically, because it is the difference between the end of the activity bar and the beginning of the bar of the succeeding activity with the earliest starting time.

In figure 52 the relationships between the activities are also shown, and this results in a "linked-bar chart". In many ways this is the ideal planning technique because it combines the analytical strength of network analysis with the clarity and ease of comprehension given by the bar chart.

Note

[1] See J. D. Parry: *The Kenyan low cost modular timber bridge,* TRRL Laboratory Report 970 (Crowthorne, Berkshire, United Kingdom, Overseas Unit, Transport and Road Research Laboratory, 1981). The bridge was designed by J. E. Collins, of the Forest Department of the Ministry of Natural Resources in Kenya, and was subsequently developed under a project sponsored by the United Nation Industrial Development Organisation (UNIDO).

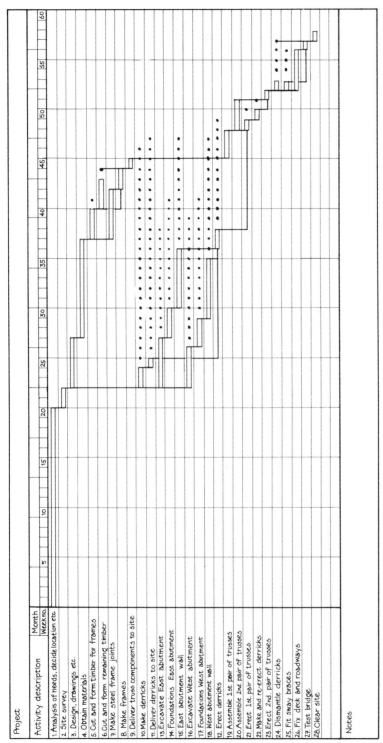

Figure 52. Bar chart from figure 50.

appendices

CHECKLISTS A

briefing stage

CLIENT

- ☐ Set up a project steering committee, if necessary.
- ☐ Appoint project manager/management team.
- ☐ Prepare project brief, time schedule and budget.
- ☐ Appoint consultants.
- ☐ Give instructions for further action.

USER

representatives of operational staff

- ☐ Consider user's requirements.
- ☐ Provide all information necessary for briefing team.
- ☐ Consider operating and maintenance factors.
- ☐ Consider occupational health and safety.
- ☐ Consider staffing, recruitment and training.
- ☐ Agree on project brief.

PROJECT MANAGEMENT TEAM

project manager and supporting services

- ☐ Establish a project organisation.
- ☐ Prepare detailed work plan and timetable for briefing.
- ☐ Advise client on choice of consultants.

- ☐ Advise client on terms of appointment for consultants.
- ☐ Determine statutory and other constraints.
- ☐ Maintain and co-ordinate progress of briefing work.
- ☐ Arrange briefing team meetings and maintain minutes.
- ☐ Pass on information to briefing team.
- ☐ Make necessary inquiries with authorities.
- ☐ Determine site ownership, boundaries and rights of way.
- ☐ Present reports with recommendations to client and obtain client's decisions.
- ☐ Prepare programme for further action.

DESIGNERS

architects, engineers, quantity surveyors and other specialists

- ☐ Examine site location and land use.
- ☐ Examine environmental implications.
- ☐ Examine user's requirements.
- ☐ Examine local planning regulations.
- ☐ Examine regulations, code, by-laws, their relationship with requirements of client/user and their implications for project cost and efficiency.
- ☐ Prepare statement of functional requirements (for civil engineering projects).
- ☐ Prepare department and room data programme (for buildings).
- ☐ Prepare floor space programme (for buildings).
- ☐ Examine site topography, including soil conditions, mining activities, access, drainage, water supply, sewerage and electricity.
- ☐ Assess type of foundation, structure, technical standards and services.
- ☐ Assess local availability of construction materials, import restrictions, means of transport and so on.

- ☐ Assess availability of labour (skilled/unskilled).
- ☐ Appraise level of local construction costs and price trends.
- ☐ Evaluate cost consequences of client's requirements.
- ☐ Prepare approximate cost estimate.
- ☐ Evaluate alternative solutions.
- ☐ Prepare sketches to illustrate alternative solutions.

PUBLIC AND OTHER AUTHORITIES

- ☐ Provide information on land-use plans, water supply, electrical connections, access roads.
- ☐ Give necessary advance approvals, permits, licences.

designing stage

CLIENT

- ☐ Make final decisions on design, site, location, format.
- ☐ Provide funds necessary for site acquisition and design work.
- ☐ Take all appropriate action for site acquisition.
- ☐ Consider and approve scheme design.
- ☐ Proceed to production and approval of tender documents.

USER

representatives of operational staff

- ☐ Provide all information necessary for design team.
- ☐ Consider operating, safety and maintenance factors.
- ☐ Consider and approve full scheme design.
- ☐ Prepare programme for recruitment and training of personnel.

PROJECT MANAGEMENT TEAM

project manager and supporting services

- ☐ Prepare detailed work plan and timetable for designing.
- ☐ Maintain and co-ordinate progress of design work.
- ☐ Arrange design team meetings and maintain minutes.
- ☐ Pass information to design team.
- ☐ Present reports with recommendations to client, and obtain decisions.
- ☐ Maintain control of costs and payments.
- ☐ Apply for all necessary permits and approvals from public authorities.
- ☐ Prepare outline construction programme and budget.

DESIGNERS

architects, engineers, quantity surveyors and other specialists

- ☐ Make investigations, analyses and other design studies.
- ☐ Prepare outline proposal.
- ☐ Prepare cost plan.
- ☐ Complete user studies and interviews.
- ☐ Complete outstanding technical studies.
- ☐ Prepare scheme design including full design of project, and preliminary design of site works, structural plans, foundations, drainage runs, design and routing of services.
- ☐ Undertake comparative cost studies.
- ☐ Provide cost checks and review cost plan.
- ☐ Carry out detail design.
- ☐ Prepare production information drawings. specifications, schedules.
- ☐ Prepare bills of quantities (if required).
- ☐ Prepare estimate of project cost.

PUBLIC AND OTHER AUTHORITIES

☐ Give final approvals: permits, licences and so on.

tendering stage

PROJECT MANAGEMENT TEAM

project manager and supporting services

☐ Obtain quotations from nominated subcontractors and suppliers for prime cost items.
☐ Prepare tender documents.
☐ Submit documents to Tender Board.
☐ If requested, assist Tender Board in the selection and approval of tenderers and in the checking and evaluation of tenders.

PUBLIC AUTHORITIES

Tender Boards

☐ Call for tenders
☐ Issue tender documents.
☐ Evaluate tenders.
☐ Award contract.
☐ Notify client and successful contractor.

CONSTRUCTION TEAM

contractor or direct labour unit, subcontractors and suppliers

☐ Prepare and submit tenders.
☐ Assess alternative construction methods, cost and time implications.
☐ Devise construction programme.
☐ Obtain information on special techniques or materials.
☐ Obtain quotations from subcontractors.

CLIENT

- ☐ Sign contract documents.
- ☐ Release funds necessary for this stage.
- ☐ Make any necessary insurance arrangements.
- ☐ Honour interim certificates.
- ☐ Note progress and approve justified increased costs.
- ☐ Honour certificate of practical completion.

USER

representatives of operational staff

- ☐ Appoint operating and maintenance staff, and undertake training.

constructing stage

PROJECT MANAGEMENT TEAM

project manager and supporting services

- ☐ Check contract documents and submit for client's and contractor's signatures.
- ☐ Examine contractor's programme and negotiate satisfactory solutions to any problems.
- ☐ Appoint site inspectorate.
- ☐ Check client's insurance.
- ☐ Arrange handover of site to contractor.
- ☐ Arrange site meetings.
- ☐ Authenticate daily work records of materials, labour and plant (only as required).
- ☐ Maintain control of final costs, client's cash flow and payments.
- ☐ Prepare regular progress reports.
- ☐ Issue interim certificates and variation orders and expedite payments.
- ☐ Issue certificate of practical completion at appropriate time.

DESIGNERS

architects, engineers, quantity surveyors and other specialists

- ☐ Provide necessary detailed production information.
- ☐ Review contractor's programme.
- ☐ Prepare programme for quality control, including special tests.
- ☐ Make regular inspections and establish (if necessary) site inspectorate.
- ☐ Prepare periodical site reports.
- ☐ Check drawings on-site in advance of work.
- ☐ Authenticate daily work records of materials, labour and plant (only as required).
- ☐ Examine and adjust priced bill of quantities.
- ☐ Prepare valuations.
- ☐ Inspect works prior to practical completion.

CONSTRUCTION TEAM

contractor or direct labour unit, subcontractors and suppliers

- ☐ Check contract documents.
- ☐ Appoint site staff and project workforce.
- ☐ Appoint subcontractors and procure building materials.
- ☐ Prepare construction programme.
- ☐ Prepare materials, plant and manpower schedules.
- ☐ Prepare cash flow budget and forecast.
- ☐ Prepare site layout.
- ☐ Arrange contractor's insurance coverage.
- ☐ Arrange production meetings.
- ☐ Direct and co-ordinate construction work, subcontractors, and deliveries.
- ☐ Collaborate in inspections prior to final completion.

PUBLIC AUTHORITIES

health boards, water boards, municipal authorities, fire authorities, and any others

☐ Make periodic checks that all statutory requirements are being observed.

commissioning stage

CLIENT

☐ Approve works when ready for take-over.
☐ Arrange insurance.
☐ Collaborate in final inspection.
☐ Honour final certificate.

USER

representatives of operational staff

☐ Take over works for occupation and operation.
☐ Receive operating manuals, as-built drawings, keys.
☐ Undertake training of operating and maintenance personnel.
☐ Report defects which require immediate action.
☐ Keep record of defects occurring during "defects liability period".
☐ Assist in final inspection.

PROJECT MANAGEMENT TEAM

project manager and supporting services

☐ Arrange hand-over meeting.
☐ Maintain and co-ordinate progress of commissioning.
☐ Issue certificate of making good defects.

- ☐ Issue certificate for release of residue of retention fund.
- ☐ Issue final certificate.
- ☐ Prepare maintenance timetables and budget.

DESIGNERS

architects, engineers, quantity surveyors and other specialists

- ☐ Inspect construction works prior to practical completion; list outstanding works.
- ☐ Take part in hand-over meeting.
- ☐ Hand over operating manuals and as-built drawings.
- ☐ Assist in training of operating and maintenance personnel.
- ☐ Inspect construction works well before end of defects liability period.
- ☐ Prepare schedule of defects.
- ☐ Prepare and agree final account.
- ☐ Undertake final inspection.

CONSTRUCTION TEAM

contractor or direct labour unit, subcontractors and suppliers

- ☐ Provide client with records for as-built drawings.
- ☐ Take part in hand-over meeting.
- ☐ Hand over keys.
- ☐ Assist in training of operating and maintenance personnel.
- ☐ Complete outstanding works.
- ☐ Start up and adjust mechanical and other systems.
- ☐ Correct defects.
- ☐ Prepare and agree final account.
- ☐ Collaborate in inspections.

PUBLIC AUTHORITIES

- ☐ Certify that statutory requirements have been satisfied.

project administration

MANAGEMENT FUNCTIONS

- ☐ Steering committee: role and responsibilities.
- ☐ Work plan.
- ☐ Organisation chart.
- ☐ Staffing.
- ☐ Job specifications for responsible staff: duties, responsibilities and authority.

PLANNING

- ☐ Time-schedules and resource plans.
- ☐ Degree of detailing during each stage.
- ☐ Planning methods.
- ☐ Responsibility for planning.

PROCUREMENT

- ☐ Procurement during each stage.
- ☐ Use of client's own resources.
- ☐ Procurement methods.
- ☐ Forms of agreement.
- ☐ Forms of payment.

CONTROL

control of work progress
- ☐ Control methods.
- ☐ Frequency of checks.

- ☐ Responsibility for control.

quality control
- ☐ Control plan.
- ☐ Instructions and variation orders.
- ☐ Inspections.

financial
- ☐ Estimates and required accuracy.
- ☐ Accounting plan.
- ☐ Project budget.
- ☐ Cost control procedures.
- ☐ Payment procedures and cash flow.
- ☐ Valuations.
- ☐ Certificates.

COMMUNICATION AND REPORTING

meetings
- ☐ Titles and purposes of meetings.
- ☐ Participants, chairman and secretary.
- ☐ Matters for discussion, standard agenda.
- ☐ Frequency, suitable days and places.
- ☐ Distribution and follow-up of minutes.
- ☐ Forms for requests for attendance, and minutes.

contacts with authorities
- ☐ List of authorities and topics.
- ☐ Routines for processing matters.
- ☐ Checklist for applications to authorities.

reports
- ☐ Nature.
- ☐ Contents.
- ☐ Frequency.
- ☐ Author, distribution.

registration, filing and distribution of documents
- ☐ List of documents.

- ☐ Numbering system.
- ☐ Registration.
- ☐ Project records.
- ☐ Distribution lists.
- ☐ Bills of delivery.
- ☐ Address list.
- ☐ Incoming, registration and distribution stamps.
- ☐ Forms.

drawings

- ☐ Scales.
- ☐ Formats.
- ☐ Layout of drawings.
- ☐ Numbering, title block.
- ☐ Classification.
- ☐ Filing.
- ☐ Registration, drawing list, distribution list, bill of delivery.
- ☐ Copying, colours, folding, number of copies.
- ☐ Extra copy orders.
- ☐ Key drawings.
- ☐ Record ("as-made") drawings.

SPECIMEN JOB DESCRIPTION B FOR PROJECT MANAGER

ORGANISATIONAL STATUS

The project manager reports directly to the client or to the project committee appointed by the client and is responsible for planning, directing and controlling the project.

DUTIES AND RESPONSIBILITIES

The duties of the project manager are—

- ☐ to establish relationships between all parties involved in the project;
- ☐ to obtain the necessary resources to carry out the work in accordance with the approved plans;
- ☐ to help to prepare the client's brief;
- ☐ to organise, instruct and supervise subordinate personnel;
- ☐ to make all necessary contacts with statutory authorities (including both inspecting and permit-granting authorities);
- ☐ to inform subordinate personnel about decisions taken;
- ☐ to set up and periodically review budgets for the project;
- ☐ to ensure that cost control takes place in conformity with established routines;
- ☐ to set up and periodically review time-schedules and resource plans for the project;
- ☐ to recommend suitable briefing design and contract procedures;

- ☐ to advise on the appointment of suitable consultants, suppliers and contractors.
- ☐ to ensure that insurances and securities are adequate and in force at all times;
- ☐ to brief the co-ordinating committee (if any) and seek its advice when necessary;
- ☐ to convene and chair project meetings, and ensure that accurate minutes are kept and distributed to all interested parties;
- ☐ to supervise the construction and defects liability period;
- ☐ to prepare periodic reports for the client on progress, cost and quality of work;
- ☐ to expedite contract payments.

AUTHORITY

The project manager is authorised—

- ☐ to use funds, personnel and other resources subject to budgets and plans approved by the client;
- ☐ to lead project personnel and set work targets for them;
- ☐ to make decisions regarding variations to contracts within limits approved by the client;
- ☐ to certify costs arising within the project;
- ☐ to represent the client in relations with ministries, consultants, contractors and suppliers.

GLOSSARY C

Accumulated payments

The running total of payments which have been made by the client.

Approvals

An acceptance by local, statutory or other authorities that the proposed construction project conforms to the regulations of those authorities.

Bill of quantities, approximate

An interim form of "bill of quantities", used to cost outline designs, which uses approximate quantities.

Bill of quantities, full

A document prepared from complete production information which states the quantities of, and describes in detail, all the building materials and components required.

Briefing stage

The initial stage in the construction process when a general outline of requirements is prepared to provide the client with proposals and recommendations, in order that he may determine the form in which the project is to proceed.

Briefing team

The team of users, designers and specialists, and in some cases the contractors and suppliers, involved during the briefing stage of the project.

Builder

See Constructor.

Capital projects

Projects that require capital investment.

Cash flow

The flow of cash funds to make the necessary cash payments during the course of a project, together with all cash received.

Certificate, final

A form which sets down the amount outstanding (including retention monies) to be paid to the contractor by the client after the final account has been agreed and any defects in materials or workmanship arising out of the works, which have emerged during the defects liability period, have been rectified.

Certificate, interim

A form which sets down the amount to be paid to the contractor by the client for work done during a specific period, usually one month. The certified amount is the result of subtracting the retention money from the valuation.

Certificate of practical completion

A form which, after a satisfactory inspection at the end of the construction phase, is issued to certify that the work has been completed. Once it has been issued, the client pays the contractor a part (usually half) of the retention money. The defect liability period then commences.

Chainage

The distance from some fixed reference point along a road or survey line.

Clerk of works

On large and many medium-sized construction projects, it is common to have a clerk of works as the resident superintendent of the works. He is responsible for checking the quality of the construction work. He may be employed either by the designer on behalf of the client, or directly by the client. In either case he is part of the project management team.

Client

The agency or individual requiring the construction project. Where the government sponsors a project, the client is usually one of its ministries.

Commissioning team

The team of users, designers, specialists, contractors and suppliers involved during the commissioning stage.

Competitive tendering

A form of tendering where invitations to submit a price for the construction of a project are sent to contractors. Stringent rules are usually attached to the related administrative procedures.

Construction team

The team of users, designers, specialists, contractors and suppliers involved during the construction stage.

Consultant

One who provides professional or expert advice.

Contract

An agreement between two or more parties.

Contract, conditions of

Any condition or prerequisite written into a contract setting out the obligations, rights and liabilities of the parties to the contract.

Contract period

The time stated in the contract for completing the construction work.

Contract, standard form of

Standardised conditions, arrangement and layout of contract documents which are generally used and accepted.

Constructor/builder

The person or private organisation (usually a contractor or direct labour force) responsible for the site construction work.

Contingencies

Allowances for costs resulting from unforeseen circumstances.

Co-ordinating committee

See Steering committee.

Cost

To the client, the cost is the price he pays the builder. To the builder, it is the price he pays for the resources used in executing the project.

Cost control

Active measures to ensure that the costs of the construction project do not exceed the project budget.

Cost diary

A record of events which have influenced project costs.

Cost forecast

An estimate of the probable final costs of the project.

Cost plus fee

A fee which includes an additional payment based on a percentage of the verified costs for a project.

Cost reimbursement contract

A form of contract in which the contractor or consultant is paid for those costs that he can show have been incurred, plus a previously agreed additional fee.

Defects liability period or maintenance period

A period following the completion of the project, during which the contractor is responsible for remedying any defects in workmanship or materials which may emerge.

Department/room data programme

A schedule outlining the functions and floor space requirements of each department or room in a building.

Design-and-construct company

A firm which undertakes the design and construction of projects, i.e. providing a single source of responsibility. Sometimes called a "package deal" or "turnkey" company.

Design(ing) team

The group responsible for design of the project, ranging in size from a single architect to a number of professionals drawn from various disciplines.

Divided contract

A contract for which separate contractors are appointed for different parts of the project, such as earthworks or foundations.

Feasibility study

A detailed investigation and analysis, conducted to determine the financial, technical or other advisability of a proposed project.

Final account

A summary of the final construction cost of the project to the client, excluding fees and interest charges. This account includes the cost of savings, and of any variations to the contract documents or alterations to provisional or prime cost items.

Final Certificate

See Certificate, final.

Final cost (or price)

The actual cost or price of the project to the client, including construction, consultants' fees, fitting-out, moving and occupational expenses, interest on monies, salaries and overheads of client staff.

Fixed price contract

A contract defining precisely the consultant's or contractor's tasks and for which a fixed price or fee, irrespective of external variations, has been negotiated.

Funds

Monetary reserves for undertaking the project.

Index costs

Index-linked allowances for costs due to price escalation caused by general inflation during the course of the project, as represented by some objective indices.

Infrastructure

The roads, sanitation, communications and water and power supplies necessary to support community life.

Instructions

Directives issued to the contractor during the construction phase of a project and resulting from variations to, or amplification of, the information contained in the contract documents.

Interim valuation

An estimate of the value of the construction work completed.

Invoice

A bill for goods or services delivered.

Liquidated damages

A sum specified in the contract to be paid to the client by the contractor if he fails to complete the works within the specified time. This sum is intended to compensate the client for damages suffered; it is not a penalty.

Lump sum

A fixed amount payment.

Minutes

A brief summary of the proposals and decisions made at a meeting, and the action to be taken.

Negotiated contract

A form of contract where a price for the construction works is negotiated by the client with a contractor.

Nominated subcontractor/supplier

Subcontractors or suppliers selected before appointment of the main contractor, or who may be so selected.

Operational estimate

A form of cost estimate based on a detailed operational plan of the construction works.

Owner

Another name for the client, usual in North American terminology.

Parastatal

Refers to a publicly owned but independent organisation or company.

Payment report

A report summarising the payments which have been made, and the present financial status of the projects with regard to the project budget.

Periodic cost forecast

An estimate of current and estimated future costs, drawn up periodically.

Prime cost items

Items of work included in bills of quantities or schedules of rates, against which a price has been placed prior to tender. These usually refer to items to be undertaken by subcontractors or provided by suppliers who have been selected, or nominated, by the client prior to tender.

Procurement

The acquisition of any kind of external resources needed to carry out the whole or part of a construction project.

Production information

Drawings, specifications, schedules and bills of quantities prepared by the design team and describing what has to be constructed.

Project budget

The sum established by the client as being available for the entire project, including land acquisition, construction, equipment, professional services, interest and contingencies.

Project management

The process of planning, executing and controlling a project from start to finish in a given time, at a given cost, for a given end product, using available human and technical resources.

Project management team

The team of specialists, such as planners, administrators and supervisors, working under the direction of a project manager, which is responsible for managing the project.

Project manager

The person with authority and responsibility to manage the project according to his terms of reference.

Provisional items

Items of work, usually given as provisional quantities or provisional sums, included in tendering documents and normally referring to items of work which may have to be carried out as part of the project but which cannot be determined with certainty before the work commences.

Quality control

Activities and methods aiming at ensuring that materials, methods, workmanship and the completed project will meet the stated requirements.

Quantity surveyor

One who estimates the amount and cost of materials and labour required for the construction of a building, and advises the client on cost matters.

Resident architect/engineer

A person often employed on large construction projects as the site representative of the architect, engineer or client.

Resource plan

A plan which summarises the availability and allocation of resources, and enables their effective scheduling.

Retention money

The money subtracted from the valuation of the work completed by the main contractor and subcontractors, and held by the client to cover the costs of remedying any defects in materials or workmanship.

Scale of fees

Approved graduated payments for various valuations of work undertaken by specialists, and published by professional bodies as a basis for ensuring quality of service.

Schedule of payments/rates

An alternative method to payment by valuation of work done. The number and value of payments are usually agreed before the construction work commences.

Sketch plans

Drawings, often free-hand, to determine the general approach to the layout, design and construction of a building.

Specification

A comprehensive description and explanation of the project, its components and materials and the required standard of workmanship.

Steering committee

An ad hoc committee which may be established by the client to direct the work and activities of the project manager.

Target cost contract

A contract in which the difference between the project's estimated cost, the target and the actual cost is shared by the client and the contractor.

Tender

An offer made by a contractor to complete the site construction work, as described in the contract documents, for a specified sum of money.

Tender documents

The set of documents on which the tenders are to be based and which are sent to the would-be tenderers. The documents usually include the project description, specifications, bills of quantities, plans and elevations and working drawings.

Tenderers

Those parties submitting replies to the invitation to tender, regarding their price and conditions.

Time plan/schedule

A time-based plan of the work to be undertaken indicating the respective order, and time for start and finish, of the activities in the project.

Two-stage tendering

A form of tendering used when the early selection of a contractor is desirable and an approximate bill of quantities is taken as the basis for price negotiations.

Type designs/drawings

Standard design proposals to a particular type of building or parts of a building.

Variations

Additions to or subtractions from the production information after the contract between the client and contractor has been signed. The additional or reduced costs of these variations are negotiated with the contractor either by the client himself or by the consultants.

Work plan

A statement indicating the sequence of work which has to be undertaken, by whom, and its appropriate timing.

Working drawings

Drawings intended for use by the contractor or subcontractor, which form part of the contract documents and provide all the necessary detailed information for site construction. Also known as production drawings.

SELECT BIBLIOGRAPHY D

M. Allal and G. A. Edmonds, in collaboration with A. S. Bhalla: *Manual on the planning of labour-intensive road construction* (Geneva, ILO, 1977).

R. E. Calvert: *Introduction to building management* (London, Newnes-Butterworths, 3rd ed., 1970).

Chartered Institute of Building: *Building for industry and commerce: Client's guide* (London, 1980).

P. P. Dharwadker: *Management in construction industry* (New Delhi, Oxford and IBH Publishing Co., 1979).

G. A. Edmonds and J. D. F. G. Howe (eds.): *Roads and resources: Appropriate technology in road construction in developing countries* (London, Intermediate Technology Publications, 1980).

C. B. Handy: *Understanding organisations* (Harmondsworth, Middlesex, Penguin Books, 1976).

D. R. Harper: *Building: The process and the product* (London, Construction Press, 1978).

F. Harris and R. McCaffer: *Modern construction management* (London, Crosby Lockwood Staples, 1977).

Her Majesty's Factory Inspectorate: *Managing safety* (London, HMSO, 1981).

ILO: *Accident prevention* (Geneva, 2nd ed., 1983).

—: *Building work: A compendium of occupational safety and health practice,* Occupational Safety and Health Series, No. 42 (Geneva, 1979).

—: *Civil engineering work: A compendium of occupational safety practice,* Occupational Safety and Health Series, No. 45 (Geneva, 1981).

—: *Encyclopaedia of occupational health and safety,* 2 vols. (Geneva, 3rd ed., 1983).

—: *Safety and health in building and civil engineering work: An ILO code of practice* (Geneva, 1972).

Institution of Civil Engineers: *Civil engineering procedure* (London, 3rd ed., 1979).

T. C. Kavanagh, F. Muller and J. J. O'Brien: *Construction management* (New York and London, McGraw-Hill, 1978).

National Economic Development Office: *Before you build: What a client needs to know about the construction industry* (London, HMSO, 1974).

Royal Institute of British Architects: *Architectural practice and management* (London, 1973).

Maurice Snowdon: *Management of engineering projects* (London, Newnes-Butterworths, 1973).